MARGOT LEE SHETTERLY is a writer who grew up in Hampton, Virginia, where she knew many of the women featured in *Hidden Figures*. The daughter of a NASA research scientist, she is an Alfred P. Sloan Foundation Fellow and the recipient of a Virginia Foundation of the Humanities grant for her research into the history of women in computing. She lives in Charlottesville, Virginia.

HIDDEN FIGURES

THE UNTOLD STORY OF
THE AFRICAN AMERICAN WOMEN
WHO HELPED WIN THE SPACE RACE

MARGOT LEE SHETTERLY

**COLLINS
MODERN
CLASSICS**

4th Estate
An imprint of HarperCollins*Publishers*
1 London Bridge Street
London SE1 9GF

4thestate.co.uk

HarperCollins*Publishers*
1st Floor, Watermarque Building, Ringsend Road
Dublin 4, Ireland

Published as a Collins Modern Classics paperback in 2022

Previously published as a William Collins paperback in 2017
First published in Great Britain by William Collins in 2016
First published in the US by William Morrow, an imprint of
HarperCollins*Publishers*, in 2016

A catalogue record for this book is available from the British Library

ISBN 978-0-00-855537-5

Printed and Bound in the UK using
100% Renewable Electricity at CPI Group (UK) Ltd

MIX
Paper from
responsible sources
FSC
www.fsc.org
FSC™ C007454

This book is produced from independently certified FSC™ paper
to ensure responsible forest management.

For more information visit: www.harpercollins.co.uk/green

To my parents, Margaret G. Lee and Robert B. Lee III,
and to all of the women at the NACA and NASA
who offered their shoulders to stand on

CONTENTS

AUTHOR'S NOTE

"Negro." "Colored." "Indian." "Girls." Though some readers might find the language of *Hidden Figures* discordant to their modern ears, I've made every attempt to remain true to the time period, and to the voices of the individuals represented in this story.

PROLOGUE

Mrs. Land worked as a computer out at Langley," my father said, taking a right turn out of the parking lot of First Baptist Church in Hampton, Virginia.

My husband and I visited my parents just after Christmas in 2010, enjoying a few days away from our full-time life and work in Mexico. They squired us around town in their twenty-year-old green minivan, my father driving, my mother in the front passenger seat, Aran and I buckled in behind like siblings. My father, gregarious as always, offered a stream of commentary that shifted fluidly from updates on the friends and neighbors we'd bumped into around town to the weather forecast to elaborate discourses on the physics underlying his latest research as a sixty-six-year-old doctoral student at Hampton University. He enjoyed touring my Maine-born-and-raised husband through our neck of the woods and refreshing my connection with local life and history in the process.

During our time home, I spent afternoons with my mother catching matinees at the local cinema, while Aran tagged along with my father and his friends to Norfolk State University football games. We gorged on fried-fish sandwiches at hole-in-the-wall joints near Buckroe Beach, visited the Hampton University Museum's Native American art collection, and haunted local antiques shops.

As a callow eighteen-year-old leaving for college, I'd seen my hometown as a mere launching pad for a life in worldlier locales, a place to be from rather than a place to be. But years and miles away from home could never attenuate the city's hold on my identity, and the more I explored places and people far from Hampton, the more my status as one of its daughters came to mean to me.

That day after church, we spent a long while catching up with the formidable Mrs. Land, who had been one of my favorite Sunday school teachers. Kathaleen Land, a retired NASA mathematician, still lived on her own well into her nineties and never missed a Sunday at church. We said our good-byes to her and clambered into the minivan, off to a family brunch. "A lot of the women around here, black and white, worked as computers," my father said, glancing at Aran in the rearview mirror but addressing us both. "Kathryn Peddrew, Ophelia Taylor, Sue Wilder," he said, ticking off a few more names. "And Katherine Johnson, who calculated the launch windows for the first astronauts."

The narrative triggered memories decades old, of spending a much-treasured day off from school at my father's office at the National Aeronautics and Space Administration's Langley Research Center. I rode shotgun in our 1970s Pontiac, my brother, Ben, and sister Lauren in the back as our father drove the twenty minutes from our house, straight over the Virgil I. Grissom Bridge, down Mercury Boulevard, to the road that led to the NASA gate. Daddy flashed his badge, and we sailed through to a campus of perfectly straight parallel streets lined from one end to the other by unremarkable two-story redbrick buildings. Only the giant hypersonic wind tunnel complex—a one-hundred-foot ridged silver sphere presiding over four sixty-foot smooth silver globes—offered visual evidence of the remarkable work occurring on an otherwise ordinary-looking campus.

Building 1236, my father's daily destination, contained a byzantine complex of government-gray cubicles, perfumed with the grown-up smells of coffee and stale cigarette smoke. His engineering colleagues with their rumpled style and distracted manner seemed like exotic birds in a sanctuary. They gave us kids stacks of discarded 11×14 continuous-form computer paper, printed on one side with cryptic arrays of numbers, the blank side a canvas for crayon masterpieces. Women occupied many of the cubicles; they answered phones and sat in front of typewriters, but they also made hieroglyphic marks on transparent slides and conferred with my father and other men in the office on the stacks of documents that littered their desks. That so many of them were African American, many of them my grandmother's age, struck me as simply a part of the natural order of things: growing up in Hampton, the face of science was brown like mine.

My dad joined Langley in 1964 as a coop student and retired in 2004 an internationally respected climate scientist. Five of my father's seven siblings made their bones as engineers or technologists, and some of his best buddies—David Woods, Elijah Kent, Weldon Staton—carved out successful engineering careers at Langley. Our next-door neighbor taught physics at Hampton University. Our church abounded with mathematicians. Supersonics experts held leadership positions in my mother's sorority, and electrical engineers sat on the board of my parents' college alumni associations. My aunt Julia's husband, Charles Foxx, was the son of Ruth Bates Harris, a career civil servant and fierce advocate for the advancement of women and minorities; in 1974, NASA appointed her deputy assistant administrator, the highest-ranking woman at the agency. The community certainly included black English professors, like my mother, as well as black doctors and dentists, black mechanics, janitors, and contractors, black cobblers, wedding planners, real estate agents, and undertakers, several black lawyers, and a handful of black Mary Kay salespeople. As a child, however, I knew so many African Americans working in science, math, and engineering that I thought that's just what black folks did.

My father, growing up during segregation, experienced a different reality. "Become a physical education teacher," my grandfather said in 1962 to his eighteen-year-old son, who was hell-bent on studying electrical engineering at historically black Norfolk State College.

In those days, college-educated African Americans with book smarts and common sense put their chips on teaching jobs or sought work at the post office. But my father, who built his first rocket in junior high metal shop class following the Sputnik launch in 1957, defied my grandfather and plunged full steam ahead into engineering. Of course, my grandfather's fears that it would be difficult for a black man to break into engineering weren't unfounded. As late as 1970, just 1 percent of all American engineers were black—a number that doubled to a whopping 2 percent by 1984. Still, the federal government was the most reliable employer of African Americans in the sciences and technology: in 1984, 8.4 percent of NASA's engineers were black.

NASA's African American employees learned to navigate their way through the space agency's engineering culture, and their successes in turn afforded their children previously unimaginable access to American society. Growing up with white friends and attending integrated schools, I took much of the groundwork they'd laid for granted.

Every day I watched my father put on a suit and back out of the driveway to make the twenty-minute drive to Building 1236, demanding the best from himself in order to give his best to the space program and to his family. Working at Langley, my father secured my family's place in the comfortable middle class, and Langley became one of the anchors of our social life. Every summer, my siblings and I saved our allowances to buy tickets to ride ponies at the annual NASA carnival. Year after year, I confided my Christmas wish list to the NASA Santa at the Langley children's Christmas party. For years, Ben, Lauren, and my youngest sister, Jocelyn, still a toddler, sat in the bleachers of the Langley Activities Building on Thursday nights, rooting for my dad and his "NBA" (NASA Basketball Association) team, the Stars. I was as much a product of NASA as the Moon landing.

The spark of curiosity soon became an all-consuming fire. I peppered my father with questions about his early days at Langley during the mid-1960s, questions I'd never asked before. The following Sunday I interviewed Mrs. Land about the early days of Langley's computing pool, when part of her job responsibility was knowing which bathroom was marked for "colored" employees. And less than a week later I was sitting on the couch in Katherine Johnson's living room, under a framed American flag that had been to the Moon, listening to a ninety-three-year-old with a memory sharper than mine recall segregated buses, years of teaching and raising a family, and working out the trajectory for John Glenn's spaceflight. I listened to Christine Darden's stories of long years spent as a data analyst, waiting for the chance to prove herself as an engineer.

Even as a professional in an integrated world, I had been the only black woman in enough drawing rooms and boardrooms to have an inkling of the chutzpah it took for an African American woman in a segregated southern workplace to tell her bosses she was sure her calculations would put a man on the Moon. These women's paths set the stage for mine; immersing myself in their stories helped me understand my own.

Even if the tale had begun and ended with the first five black women who went to work at Langley's segregated west side in May 1943—the women later known as the "West Computers"—I still would have committed myself to recording the facts and circumstances of their lives. Just as islands—isolated places with unique, rich biodiversity—have relevance for the ecosystems everywhere, so does studying seemingly isolated or overlooked people and events from the past turn up unexpected connections and insights to modern life. The idea that black women had been recruited to work as mathematicians at the NASA installation in the South during the days of segregation defies our expectations and challenges much of what we think we know about American history. It's a great story, and that alone makes it worth telling.

In the early stages of researching this book, I shared details of what I had found with experts on the history of the space agency. To a person they encouraged what they viewed as a valuable addition to the body of knowledge, though some questioned the magnitude of the story.

"How many women are we talking about? Five or six?"

I had known more than that number just growing up in Hampton, but even I was surprised at how the numbers kept adding up. These women showed up in photos and phone books, in sources both expected and unusual. A mention of a Langley job in an engagement announcement in the *Norfolk Journal and Guide*. A handful of names from the daughter of one of the first West Computers. A 1951 memo from the Langley personnel officer reporting on the numbers and status of its black employees, which unexpectedly made reference to one black woman who was a "GS-9 Research Scientist." I discovered one 1945 personnel document describing a beehive of mathematical activity in an office in a new building on Langley's west side, staffed by twenty-five black women coaxing numbers out of calculators on a twenty-four-hour schedule, overseen by three black shift supervisors who reported to two white head computers. Even as I write the final words of this book, I'm still doing the numbers. I can put names to almost fifty black women who worked as computers, mathematicians, engineers, or scientists at the Langley Memorial Aeronautical Laboratory from 1943 through 1980, and my intuition is that twenty more names can be shaken loose from the archives with more research.

And while the black women are the most hidden of the mathematicians who worked at the NACA, the National Advisory Committee for Aeronautics, and later at NASA, they were not sitting alone in the shadows: the white women who made up the majority of Langley's computing workforce over the years have hardly been recognized for their contributions to the agency's long-term success. Virginia Biggins worked the Langley beat for the *Daily Press* newspaper, covering the space program starting in 1958. "Everyone said, 'This is a scientist, this is an engineer,' and it was always a man," she said in a 1990 panel on Langley's human computers. She never got to meet any of the women. "I just assumed they were all secretaries," she said. Five white women joined Langley's first computing pool in 1935, and by 1946, four hundred "girls" had already been trained as aeronautical foot soldiers. Historian Beverly Golemba, in a 1994 study, estimated that Langley had employed "several hundred" women as human computers. On the tail end of the research for *Hidden Figures*, I can now see how that number might top one thousand.

To a first-time author with no background as a historian, the stakes involved in writing about a topic that was virtually absent from the history books felt high. I'm sensitive to the cognitive dissonance conjured by the phrase "black female mathematicians at NASA." From the beginning, I knew that I would have to apply the same kind of analytical reasoning to my research that these women applied to theirs. Because as exciting as it was to discover name after name, finding out who they were was just the first step. The real challenge was to document their work. Even more than the surprisingly large numbers of black and white women who had been hiding in a profession seen as universally white and male, the body of work they left behind was a revelation.

There was Dorothy Hoover, working for Robert T. Jones in 1946 and publishing theoretical research on his famed triangle-shaped delta wings in 1951. There was Dorothy Vaughan, working with the white "East Computers" to write a textbook on algebraic methods for the mechanical calculating machines that were their constant companions. There was Mary Jackson, defending her analysis against John Becker, one of the world's top aerodynamicists. There was Katherine Coleman Goble Johnson, describing the orbital trajectory of John Glenn's flight, the math in her trailblazing 1959 report as elegant and precise and grand as a symphony. There was Marge Hannah, the white computer who served as the black women's first boss, coauthoring a report with Sam Katzoff, who became the laboratory's chief scientist. There was Doris Cohen, setting the bar for them all with her first research report—the NACA's first female author—back in 1941.

My investigation became more like an obsession; I would walk any trail if it meant finding a trace of one of the computers at its end. I was determined to prove their existence and their talent in a way that meant they would never again be lost to history. As the photos and memos and equations and family stories became real people, as the women became my companions and returned to youth or returned to life, I started to want something more for them than just putting them on the record. What I wanted was for them to have the grand, sweeping narrative that they deserved, the kind of American history that belongs to the Wright Brothers and the astronauts, to Alexander Hamilton and Martin Luther King Jr. Not told as a separate history, but as a part of the story we all know. Not at the margins, but at the very center, the protagonists of the drama. And not just because they are black, or because they are women, but because they are part of the American epic.

Today, my hometown—the hamlet that in 1962 dubbed itself "Spacetown USA"—looks like any suburban city in a modern and hyperconnected America. People of all races and nationalities mingle on Hampton's beaches and in its bus stations, the WHITES ONLY signs of the past now relegated to the local history museum and the memories of survivors of the civil rights revolution. Mercury Boulevard no longer conjures images of the eponymous mission that shot the first Americans beyond the atmosphere, and each day the memory of Virgil Grissom fades away from the bridge that bears his name. A downsized space program and decades of government cutbacks have hit the region hard; today, an ambitious college grad with a knack for numbers might set her sights on a gig at a Silicon Valley startup or make for one of the many technology firms that are conquering the NASDAQ from the Virginia suburbs outside of Washington, DC.

But before a computer became an inanimate object, and before Mission Control landed in Houston; before Sputnik changed the course of history, and before the NACA became NASA; before the Supreme Court case *Brown v. Board of Education of Topeka* established that separate was in fact not equal, and before the poetry of Martin Luther King Jr.'s "I Have a Dream" speech rang out over the steps of the Lincoln Memorial, Langley's West Computers were helping America dominate aeronautics, space research, and computer technology, carving out a place for themselves as female mathematicians who were also black, black mathematicians who were also female. For a group of bright and ambitious African American women, diligently prepared for a mathematical career and eager for a crack at the big leagues, Hampton, Virginia, must have felt like the center of the universe.

HIDDEN
FIGURES

CHAPTER ONE

A Door Opens

Melvin Butler, the personnel officer at the Langley Memorial Aeronautical Laboratory, had a problem, the scope and nature of which was made plain in a May 1943 telegram to the civil service's chief of field operations. "This establishment has urgent need for approximately 100 Junior Physicists and Mathematicians, 100 Assistant Computers, 75 Minor Laboratory Apprentices, 125 Helper Trainees, 50 Stenographers and Typists," exclaimed the missive. Every morning at 7:00 a.m., the bow-tied Butler and his staff sprang to life, dispatching the lab's station wagon to the local rail depot, the bus station, and the ferry terminal to collect the men and women—so many women now, each day more women—who had made their way to the lonely finger of land on the Virginia coast. The shuttle conveyed the recruits to the door of the laboratory's Service Building on the campus of Langley Field. Upstairs, Butler's staff whisked them through the first-day stations: forms, photos, and the oath of office: *I will support and defend the Constitution of the United States against all enemies, foreign and dometic . . . so help me God.*

Thus installed, the newly minted civil servants fanned out to take their places in one of the research facility's expanding inventory of buildings, each already as full as a pod ripe with peas. No sooner had

Sherwood Butler, the laboratory's head of procurement, set the final brick on a new building than his brother, Melvin, set about filling it with new employees. Closets and hallways, stockrooms and workshops stood in as makeshift offices. Someone came up with the bright idea of putting two desks head to head and jury-rigged the new piece of furniture with a jump seat in order to squeeze three workers into space designed for two. In the four years since Hitler's troops overran Poland—since American interests and the European war converged in an all-consuming conflict—the laboratory's complement of 500-odd employees at the close of the decade was on its way to 1,500. Yet the great groaning war machine swallowed them whole and remained hungry for more.

The offices of the Administration Building looked out upon the crescent-shaped airfield. Only the flow of civilian-clothed people heading to the laboratory, the oldest outpost of the National Advisory Committee for Aeronautics (NACA), distinguished the low brick buildings belonging to that agency from identical ones used by the US Army Air Corps. The two installations had grown up together, the air base devoted to the development of America's military airpower capability, the laboratory a civilian agency charged with advancing the scientific understanding of aeronautics and disseminating its findings to the military and private industry. Since the beginning, the army had allowed the laboratory to operate on the campus of the airfield. The close relationship with the army flyers served as a constant reminder to the engineers that every experiment they conducted had real-world implications.

The double hangar—two 110-foot-long buildings standing side by side—had been covered in camouflage paint in 1942 to deceive enemy eyes in search of targets, its shady and cavernous interior sheltering the machines and their minders from the elements. Men in canvas jumpsuits, often in groups, moved in trucks and jeeps from plane to plane, stopping to hover at this one or that like pollinating insects, checking them, filling them with gas, replacing parts, examining them, becoming one with them and taking off for the heavens. The music of airplane engines and propellers cycling through the various movements of takeoff, flight, and landing played from before sunrise until dusk, each machine's sounds as unique to its minders as a baby's cry to its mother. Beneath the tenor notes of the engines played the bass roar of the laboratory's wind tunnels, turning their on-demand hurricanes onto the planes—plane parts, model planes, full-sized planes.

Just two years prior, with the storm clouds gathering, President Roosevelt challenged the nation to ramp up its production of airplanes to fifty thousand per year. It seemed an impossible task for an industry that as recently as 1938 had only provided the Army Air Corps with ninety planes a month. Now, America's aircraft industry was a production miracle, easily surpassing Roosevelt's mark by more than half. It had become the largest industry in the world, the most productive, the most sophisticated, outproducing the Germans by more than three times and the Japanese by nearly five. The facts were clear to all belligerents: the final conquest would come from the sky.

For the flyboys of the air corps, airplanes were mechanisms for transporting troops and supplies to combat zones, armed wings for pursuing enemies, sky-high launching pads for ship-sinking bombs. They reviewed their vehicles in an exhaustive preflight checkout before climbing into the sky. Mechanics rolled up their sleeves and sharpened their eyes; a broken piston, an improperly locked shoulder harness, a faulty fuel tank light, any one of these could cost lives. But even before the plane responded to its pilot's knowing caress, its nature, its very DNA—from the shape of its wings to the cowling of its engine—had been manipulated, refined, massaged, deconstructed, and recombined by the engineers next door.

Long before America's aircraft manufacturers placed one of their newly conceived flying machines into production, they sent a working prototype to the Langley laboratory so that the design could be tested and improved. Nearly every high-performance aircraft model the United States produced made its way to the lab for drag cleanup:

the engineers parked the planes in the wind tunnels, making note of air-disturbing surfaces, bloated fuselages, uneven wing geometries. As prudent and thorough as old family doctors, they examined every aspect of the air flowing over the plane, making careful note of the vital signs. NACA test pilots, sometimes with an engineer riding shotgun, took the plane for a flight. Did it roll unexpectedly? Did it stall? Was it hard to maneuver, resisting the pilot like a shopping cart with a bad wheel? The engineers subjected the airplanes to tests, capturing and analyzing the numbers, recommending improvements, some slight, others significant. Even small improvements in speed and efficiency multiplied over millions of pilot miles added up to a difference that could tip the long-term balance of the war in the Allies' favor.

"Victory through airpower!" Henry Reid, engineer-in-charge of the Langley laboratory, crooned to his employees, the shibboleth a reminder of the importance of the airplane to the war's outcome. "Victory through airpower!" the NACA-ites repeated to each other, minding each decimal point, poring over differential equations and pressure distribution charts until their eyes tired. In the battle of research, victory would be theirs.

Unless, of course, Melvin Butler failed to feed the three-shift-a-day, six-day-a-week operation with fresh minds. The engineers were one thing, but each engineer required the support of a number of others: craftsmen to build the airplane models tested in the tunnels, mechanics to maintain the tunnels, and nimble number crunchers to process the numerical deluge that issued from the research. Lift and drag, friction and flow. What was a plane but a bundle of physics? Physics, of course, meant math, and math meant mathematicians. And since the middle of the last decade, mathematicians had meant women. Langley's first female computing pool, started in 1935, had caused an uproar among the men of the laboratory. How could a female mind process something so rigorous and precise as math? The very idea, investing $500 on a calculating machine so it could be used by a girl! But the "girls" had been good, very good—better at computing, in fact, than many of the engineers, the men themselves grudgingly admitted. With only a handful of girls winning the title "mathematician"—a professional designation that put them on equal footing with entry-level male employees—the fact that most computers were designated as lower-paid "subprofessionals" provided a boost to the laboratory's bottom line.

But in 1943, the girls were harder to come by. Virginia Tucker, Langley's head computer, ran laps up and down the East Coast searching for coeds with even a modicum of analytical or mechanical skill, hoping for matriculating college students to fill the hundreds of open positions for computers, scientific aides, model makers, laboratory assistants, and yes, even mathematicians. She conscripted what seemed like entire classes of math graduates from her North Carolina alma mater, the Greensboro College for Women, and hunted at Virginia schools like Sweetbriar in Lynchburg and the State Teachers College in Farmville.

Melvin Butler leaned on the US Civil Service Commission and the War Manpower Commission as hard as he could so that the laboratory might get top priority on the limited pool of qualified applicants. He penned ads for the local newspaper, the *Daily Press*: "Reduce your household duties! Women who are not afraid to roll up their sleeves and do jobs previously filled by men should call the Langley Memorial Aeronautical Laboratory," read one notice. Fervent pleas from the personnel department were published in the employee newsletter *Air Scoop*: "Are there members of your family or others you know who would like to play a part in gaining supremacy of the air? Have you friends of either sex who would like to do important work toward winning and shortening the war?" With men being absorbed into the military services, with women already in demand by eager employers, the labor market was as exhausted as the war workers themselves.

A bright spot presented itself in the form of another man's problem. A. Philip Randolph, the head of the largest black labor union in the country, demanded that Roosevelt open lucrative war jobs to Negro applicants, threatening in the summer of 1941 to bring one hundred thousand Negroes to the nation's capital in protest if the president rebuffed his demand. "Who the hell is this guy Randolph?" fumed Joseph Rauh, the president's aide. Roosevelt blinked.

A "tall courtly black man with Shakespearean diction and the stare of an eagle," Asa Philip Randolph, close friend of Eleanor Roosevelt, headed the 35,000-strong Brotherhood of Sleeping Car Porters. The porters waited on passengers in the nation's segregated trains, daily enduring prejudice and humiliation from whites. Nevertheless, these jobs were coveted in the black community because they provided a measure of economic stability and social standing. Believing that civil rights were inextricably linked to economic rights, Randolph fought tirelessly for the right of Negro Americans to participate fairly in the wealth of the country they had helped build. Twenty years in the future, Randolph would address the multitudes at another March on Washington, then concede the stage to a young, charismatic minister from Atlanta named Martin Luther King Jr.

Later generations would associate the black freedom movement with King's name, but in 1941, as the United States oriented every aspect of its society toward war for the second time in less than thirty years, it was Randolph's long-term vision and the specter of a march that never happened that pried open the door that had been closed like a bank vault since the end of Reconstruction. With two strokes of a pen—Executive Order 8802, ordering the desegregation of the defense industry, and Executive Order 9346, creating the Fair Employment Practices Committee to monitor the national project of economic inclusion—Roosevelt primed the pump for a new source of labor to come into the tight production process.

Nearly two years after Randolph's 1941 showdown, as the laboratory's personnel requests reached the civil service, applications of qualified Negro female candidates began filtering in to the Langley Service Building, presenting themselves for consideration by the laboratory's personnel staff. No photo advised as to the applicant's color—that requirement, instituted under the administration of Woodrow Wilson, was struck down as the Roosevelt administration tried to dismantle discrimination in hiring practices. But the applicants' alma maters tipped their hand: West Virginia State University, Howard, Arkansas Agricultural, Mechanical & Normal, Hampton Institute just across town—all Negro schools. Nothing in the applications indicated anything less than fitness for the job. If anything, they came with more experience than the white women applicants, with many years of teaching experience on top of math or science degrees.

They would need a separate space, Melvin Butler knew. Then they would have to appoint someone to head the new group, an experienced girl—white, obviously—someone whose disposition suited the sensitivity of the assignment. The Warehouse Building, a brand-new space on the west side of the laboratory, a part of the campus that was still more wilderness than anything resembling a workplace, could be just the thing. His brother Sherwood's group had already moved there, as had some of the employees in the personnel department. With round-the-clock pressure to test the airplanes queued up in the hangar, engineers would welcome the additional hands. So many of the engineers were Northerners, relatively agnostic on the racial issue but devout when it came to mathematical talent.

Melvin Butler himself hailed from Portsmouth, just across the bay from Hampton. It required no imagination on his part to guess what some of his fellow Virginians might think of the idea of integrating Negro women into Langley's offices, the "come-heres" (as the Virginians called the newcomers to the state) and their strange ways be damned. There had always been Negro employees in the lab—janitors, cafeteria workers, mechanic's assistants, groundskeepers. But opening the door to Negroes who would be professional peers, that was something new.

Butler proceeded with discretion: no big announcement in the *Daily Press*, no fanfare in *Air Scoop*. But he also proceeded with direction: nothing to herald the arrival of the Negro women to the laboratory, but nothing to derail their arrival either. Maybe Melvin Butler was progressive for his time and place, or maybe he was just a functionary carrying out his duty. Maybe he was both. State law—and Virginia custom—

kept him from truly progressive action, but perhaps the promise of a segregated office was just the cover he needed to get the black women in the door, a Trojan horse of segregation opening the door to integration. Whatever his personal feelings on race, one thing was clear: Butler was a Langley man through and through, loyal to the laboratory, to its mission, to its worldview, and to its charge during the war. By nature—and by mandate—he and the rest of the NACA were all about practical solutions.

So, too, was A. Philip Randolph. The leader's indefatigable activism, unrelenting pressure, and superior organizing skills laid the foundation for what, in the 1960s, would come to be known as the civil rights movement. But there was no way that Randolph, or the men at the laboratory, or anyone else could have predicted that the hiring of a group of black female mathematicians at the Langley Memorial Aeronautical Laboratory would end at the Moon.

Still shrouded from view were the great aeronautical advances that would crush the notion that faster-than-sound flight was a physical impossibility, the electronic calculating devices that would amplify the power of science and technology to unthinkable dimensions. No one anticipated that millions of wartime women would refuse to leave the American workplace and forever change the meaning of women's work, or that American Negroes would persist in their demands for full access to the founding ideals of their country and not be moved. The black female mathematicians who walked into Langley in 1943 would find themselves at the intersection of these great transformations, their sharp minds and ambitions contributing to what the United States would consider one of its greatest victories.

But in 1943, America existed in the urgent present. Responding to the needs of the here and now, Butler took the next step, making a note to add another item to Sherwood's seemingly endless requisition list: a metal bathroom sign bearing the words COLORED GIRLS.

CHAPTER TWO

Mobilization

There was no escaping the heat in the summer of 1943, not in the roiling seas of the South Pacific, not in the burning skies over Hamburg and Sicily, and not for the group of Negro women working in Camp Pickett's laundry boiler plant. The temperature and humidity inside the army facility were so intense that slipping outdoors into the 100-plus degrees of the central Virginia June summer invited relief.

The laundry room was both one of the war's obscure crannies and a microcosm of the war itself, a sophisticated, efficient machine capable of processing eighteen thousand bundles of laundry each week. One group of women loaded soiled laundry into the enormous boilers. Others heaved the sopping clothes into the dryers. Another team worked the pressing machines like cooks at a giant griddle. Thirty-two-year-old Dorothy Vaughan stood at the sorting station, reuniting wayward socks and trousers with the laundry bags of the black and white soldiers who came to Camp Pickett by the trainload for four weeks of basic training before heading on to the Port of Embarkation in Newport News. Small talk of husbands, children, lives back home, or the ever-present war rose above the thunder and hum of the giant laundry boilers and dryers. *We gave him a real nice send-off, whole neighborhood turned out. Just as well you can't get stockings nowhere, hot as it is. That Mr. Randolph sure is something, and friends with Mrs. Roosevelt too!* They brooded over the husbands and brothers and fathers heading into the conflict that was so far away from the daily urgencies of their lives in Virginia, yet so close to their prayers and their dreams.

The majority of the women who found their way to the military laundry room had left behind jobs as domestic servants or as stemmers in the tobacco factories. The laundry was a humid inferno, the work as monotonous as it was uncomfortable. Laundry workers existed at the bottom of the war's great pyramid, invisible and invaluable at the same time. One aircraft industry executive estimated that each laundry worker supported three workers at his plants; with someone else to tend to their dirty clothing, men and women on the production lines had lower rates of absenteeism. The laundry workers earned 40 cents an hour, ranking them among the lowest paid of all war workers, but with few job options available to them, it felt like a windfall.

Only a week had elapsed between the end of the school year at Robert Russa Moton, the Negro high school in Farmville, Virginia, where Dorothy worked as a math teacher, and her first day of work at Camp Pickett. As a college graduate and a teacher, she stood near the top of what most Negro women could hope to achieve. Teachers were considered the "upper level of training and intelligence in the race," a ground force of educators who would not just impart book learning but live in the Negro community and "direct its thoughts and head its social movements." Her in-laws were mainstays of the town's Negro elite. They owned a barbershop, a pool hall, and a service station. The family's activities were regular fodder for the social column in the Farmville section of the *Norfolk Journal and Guide*, the leading Negro newspaper in the southeastern United States. Dorothy, her husband, Howard, and their four young children lived in a large, rambling Victorian house on South Main Street with Howard's parents and grandparents.

In the summer of 1943, Dorothy jumped at the chance to head to Camp Pickett and earn extra money during the school break. Though teaching offered prestige, the compensation was modest. Nationally,

Virginia's white teachers ranked in the bottom quarter in public school salaries, and their black counterparts might earn almost 50 percent less. Many black teachers in the South gave lessons in one- or two-room schools that barely qualified as buildings. Teachers were called upon to do whatever was necessary to keep the schoolhouses clean, safe, and comfortable for pupils. They shoveled coal in winters, fixed broken windows, scrubbed dirty floors, and prepared lunch. They reached into their own threadbare purses when the schoolroom kitty fell short.

Another woman in Dorothy's situation might have seen taking the laundry job as unthinkable, regardless of the economics. Wasn't the purpose of a college degree to get away from the need to work dirty and difficult jobs? And the location of the camp, thirty miles southeast of Farmville, meant that she lived in worker housing during the week and got back home only on weekends. But the 40 cents an hour Dorothy earned as a laundry sorter bested what she earned as a teacher, and with four children, a summer of extra income would be put to good and immediate use.

And Dorothy was of an unusually independent mind, impatient with the pretensions that sometimes accompanied the upwardly mobile members of the race. She did nothing to draw attention to herself at Camp Pickett, nor did she make any distinctions between herself and the other women. There was something in her bearing that transcended her soft voice and diminutive stature. Her eyes dominated her lovely, caramel-hued face—almond-shaped, wide-set, intense eyes that seemed to see everything. Education topped her list of ideals; it was the surest hedge against a world that would require more of her children than white children, and attempt to give them less in return. The Negro's ladder to the American dream was missing rungs, with even the most outwardly successful blacks worried that at any moment the forces of discrimination could lay waste to their economic security. Ideals without practical solutions were empty promises. Standing on her feet all day in the sweltering laundry was an opportunity if the tumbled military uniforms bought new school clothes, if each sock made a down payment on her children's college educations.

At night in the bunk of the workers' housing, as she willed a breeze to cut through the motionless night air, Dorothy thought of Ann, age eight, Maida, six, Leonard, three, and Kenneth, just eight months old. Their lives and futures informed every decision she made. Like virtually every Negro woman she knew, she struggled to find the balance between spending time with her children at home and spending time *for* them, for her family, at a job.

Dorothy was born in 1910 in Kansas City, Missouri. Her own mother died when Dorothy was just two years old, and less than a year later, her father, Leonard Johnson, a waiter, remarried. Her stepmother, Susie Peeler Johnson, worked as a charwoman at the grand Union Station train depot to help support the family. She took Dorothy as her own daughter and pushed her to succeed, teaching the precocious girl to read before she entered school, which vaulted her ahead two grades. She also encouraged her daughter's natural musical talent by enrolling her in piano lessons. When Dorothy was eight, the family relocated to Morgantown, West Virginia, where her father accepted a job working for a successful Negro restaurateur. There she attended the Beechhurst School, a consolidated Negro school located around the corner from West Virginia University, the state's flagship white college. Seven years later, Dorothy reaped the reward for her hard work in the form of the valedictorian's spot and a full-tuition scholarship to Wilberforce University, the country's oldest private Negro college, in Xenia, Ohio. The African Methodist Episcopal Sunday School Convention of West Virginia, which underwrote the scholarship, celebrated fifteen-year-old Dorothy in an eight-page pamphlet that it published and distributed to church members, lauding her intelligence, her work ethic, her naturally kind disposition, and her humility. "This is the dawn of a life, a promise held forth. We who have been fortunate enough to guide that genius and help mold it, even for a little while, will look on with interest during the coming years," wrote Dewey Fox, the organization's vice president. Dorothy was the kind of young person who filled the Negro race with hope that its future in America would be more propitious than its past.

At Wilberforce, Dorothy earned "splendid grades" and chose math as her major. When she was an upperclassman, one of Dorothy's professors at Wilberforce recommended her for graduate study in mathematics at Howard University, in what would be the inaugural class for a master's degree in the subject. Howard, based in Washington, DC, was the summit of Negro scholarship. Elbert Frank Cox and Dudley Weldon Woodard, the first two Negroes to earn doctorates in mathematics, with degrees from Cornell and the University of Pennsylvania, respectively, ran the department. The white schools' prejudice was the black schools' windfall: with almost no possibility of securing a faculty position at a white college, brilliant black scholars like Cox and Woodard and W. E. B. Du Bois, the sociologist and historian who was the first Negro to receive a doctorate from Harvard, taught almost exclusively at Negro schools, bringing students like Dorothy into close contact with some of the finest minds in the world.

Howard University represented a singular opportunity for Dorothy, in line with the AME scholarship committee's lofty expectations. Possessed of an inner confidence that attributed no shortcoming either to her race or to her gender, Dorothy welcomed the chance to prove herself in a competitive academic arena. But the economic reality that confronted Dorothy when she came out of college made graduate study seem like an irresponsible extravagance. With the onset of the Great Depression, Dorothy's parents, like a third of all Americans, found steady work hard to come by. An extra income would help keep the household above water and improve the odds that Dorothy's sister might be able to follow her path to college. Dorothy, though only nineteen years old, felt it was her responsibility to ensure that the family could make its way through the hard times, even though it meant closing the door on her own ambitions, at least for the moment. She opted to earn a degree in education and pursue teaching, the most stable career for a black woman with a college degree.

Through an extensive grapevine, black colleges received calls from schools around the country requesting teachers, then dispatched their alumni to fill open positions in everything from tar paper shacks in the rural cotton belt to Washington, DC's elite Dunbar High School. New educators hoped to teach in their major subject, of course, but would be expected to assume whatever duties were necessary. After graduation in 1929, Dorothy was sent forth like a secular missionary to join the Negro teaching force.

Her first job, teaching math and English at a Negro school in rural Tamms, Illinois, ended after her first school year. The Depression-fueled collapse in cotton prices hit the area hard, and the school system simply shut its doors, leaving no public education for the rural county's Negro students. She fared no better in her next posting in coastal North Carolina, where, in the middle of the school year, the school ran out of money and simply stopped paying her. Dorothy supported herself and contributed to the family by working as a waitress at a hotel in Richmond, Virginia, until 1931, when she got word of a job at the school in Farmville.

It was no surprise that the newcomer with the beautiful eyes caught the attention of one of Farmville's most eligible bachelors. Tall, charismatic, and quick with a smile, Howard Vaughan worked as an itinerant bellman at luxury hotels, going south to Florida in the winter and north to upstate New York and Vermont in the summer. Some years he found work closer to home at the Greenbrier, the luxury resort in White Sulphur Springs, West Virginia, which was a destination for wealthy and fabulous people from around the world.

Though her husband's work kept him on the road, Dorothy exchanged her traveling shoes for Farmville life and the routines of family, the stability of regular work, and community. Still, coming of age and entering the workforce in the depths of the Depression permanently affected Dorothy's worldview. She dressed plainly and modestly, spurned every extravagance, and never turned down the chance to put money in the bank. Though she was a member of Farmville's Beulah AME Church, it was the First Baptist Church that enjoyed her esteemed piano playing come Sunday morning, because they had hired her as their pianist.

•   •   •

As the war intensified, the town post office was awash in civil service job bulletins, competing for the eyes of locals and college students alike. It was on a trip to the post office during the spring of 1943 that Dorothy spied the notice for the laundry job at Camp Pickett. But the word on another bulletin also caught her eye: mathematics. A federal agency in Hampton sought women to fill a number of mathematical jobs having to do with airplanes. The bulletin, the handiwork of Melvin Butler and the NACA personnel department, was most certainly meant for the eyes of the white, well-to-do students at the all-female State Teachers College there in Farmville. The laboratory had sent application forms, civil service examination notices, and booklets describing the NACA's work to the school's job placement offices, asking faculty and staff to spread the word about the open positions among potential candidates. "This organization is considering a plan to visit certain women's colleges in this area and interview senior students majoring in mathematics," the laboratory wrote. "It is expected that outstanding students will be offered positions in this laboratory." Interviews that year yielded four new Farmville girls for the laboratory's computing sections.

Dorothy's house on South Main sat down the street from the college campus. Every morning as she walked the two blocks to her job at Moton High School, a U-shaped building perched on a triangular block at the south end of town, she saw the State Teachers College coeds with their books, disappearing into classrooms in their leafy sanctuary of a campus. Dorothy walked to school on the other side of the street, toeing the invisible line that separated them.

It would no sooner have occurred to her that a place with so baroque a name as the Langley Memorial Aeronautical Laboratory would solicit an application from Negro women than that the white women at the college across the street would beckon her through the front doors of their manicured enclave. Black newspapers, however, worked relentlessly to spread the word far and wide about available war jobs and exhorted their readers to apply. Some were dubbing Executive Order 8802 and the Fair Employment Practices Committee "the most significant move on the part of the Government since the Emancipation

Proclamation." Dorothy's own sister-in-law had moved to Washington to take a job in the War Department.

In the first week of May 1943, the *Norfolk Journal and Guide* published an article that would call to Dorothy like a signpost for the road not taken. "Paving the Way for Women Engineers," read the headline. The accompanying photo showed eleven well-dressed Negro women in front of Hampton Institute's Bemis Laboratory, graduates of Engineering for Women, a war training class. Founded in 1868, Hampton Institute had grown out of the classes held by the free Negro teacher Mary Peake, in the shade of a majestic tree known as the Emancipation Oak. On the eve of World War II, Hampton was one of the leading Negro colleges in the country and the focal point of the black community's participation in the conflict.

The women had come from points up and down the East Coast, and from right there in town. Pearl Bassette, one of several Hampton natives, was the daughter of a well-known black lawyer, her family tracing its roots back to the early days of the city. Ophelia Taylor, originally from Georgia, graduated from Hampton Institute, and prior to starting the class was running a nursery school. Mary Cherry came from North Carolina, Minnie McGraw from South Carolina, Madelon Glenn from faraway Connecticut. Miriam Mann, a tiny firebrand who had taught school in Georgia, had come to the city with her family when her husband, William, accepted a position as an instructor teaching machine shop at the US Naval Training School at Hampton Institute.

There were black jobs, and there were *good* black jobs. Sorting in the laundry, making beds in white folks' houses, stemming in the tobacco plant—those were black jobs. Owning a barbershop or a funeral home, working in the post office, or riding the rails as a Pullman porter—those were *good* black jobs. Teacher, preacher, doctor, lawyer—now those were *very* good black jobs, bringing stability and the esteem that accompanied formal training.

But the job at the aeronautical laboratory was something new, something so unusual it hadn't yet entered the collective dreams. Not even the long-stalled plan to equalize Negro teachers' salaries with those of their white counterparts could beat this opportunity. Even if the war ended in six months or a year, a much higher salary even for that brief time would bring Dorothy that much closer to assuring her children's future.

So that spring, Dorothy Vaughan carefully filled out and mailed two job applications: one to work at Camp Pickett, where the need for labor was so great, so undifferentiated, that there was virtually no possibility that they would not hire her. The other, much longer application reviewed her qualifications in detail. Work history. Personal references. Schools attended: high school and college. Courses taken, grades received. Languages spoken (French, which she had studied at Wilberforce). Foreign travels (None). *Would you be willing to accept a position abroad?* (No). *Would you be willing to accept a position in Washington, DC?* (Yes). *How soon could you be ready to start work?* She knew the answer before her fingers carved it into the blank: 48 hours, she wrote. I can be ready to go within forty-eight hours.

CHAPTER THREE

Past Is Prologue

The 1943 school year at Farmville's Robert Russa Moton High School started the same way other years always had: same space, more students. The "new" high school, built in 1939 to accommodate 180 students, had been obsolete almost from its beginning. In the school's first year of operation, 167 students arrived for classes. Four years later, Dorothy Vaughan and her twelve fellow teachers were welcoming 301 education-hungry youngsters, urged along by parents who wanted more for their children than a life of work in the tobacco factories. The students walked for miles to get to the school or took their chances each morning in barely roadworthy buses that made the rounds in the outer reaches of Prince Edward County.

As a member of Moton's parent-teacher association and a founding board member of the Farmville chapter of the NAACP, Dorothy worked hard to improve the long-term educational prospects of the young people of Farmville. As a teacher, her ambitions were more immediate: with only eight classrooms; no gymnasium, lockers, or cafeteria; and an auditorium outfitted with folding chairs, it took all her leadership and creativity to maintain an orderly learning environment. Somehow, she managed to impart the finer points of arithmetic and algebra in the auditorium, with two other classes taking place simultaneously. The school building might have been modest, but Dorothy's standards were not. She once discovered an error in one of the math textbooks she used in her classroom and dashed off a letter to the publisher informing them of their mistake (they fixed it, and sent her a thank-you letter in return). The Good Lord himself might have squirmed in his seat if Mrs. Vaughan had caught Him out in her class without having done His algebra homework. She devoted time after the end of the school day to tutoring students who required extra help. She also worked with the school choir; under her direction, several of Moton's vocal quartets had come away victorious in statewide music competitions. In 1935, a *Norfolk Journal and Guide* article covering the annual event dubbed her "the festival's most enthusiastic and hardest working director." In 1943, she and the school's music teacher, Altona Johns, put students through their paces in preparation for the year's Christmas cantata, "The Light Still Shines."

The feverish summer gave way to fall foliage and brisk mornings, but routines had changed to accommodate the war. The school's 4-H club made care packages for departing servicemen and hosted a community discussion entitled "What Can We Do to Win the War?" The Moton school office put war stamps on sale, each purchase a small offset against the gargantuan cost of the military production. The community held going-away parties and prepared feasts for the young men heading off to the front. Dorothy updated her classes with a unit called Wartime Mathematics, teaching students to apply arithmetic operations to household budgeting and wartime ration books and updating classic word problems with airplanes instead of cars.

Sometimes, it seemed as if Dorothy had never been without Farmville or Farmville without her. The town had embraced her with the warmth accorded a native daughter; she had called it home longer than any other place she'd lived in in her thirty-two years. Her life, however, was a model of America's great love affair with mobility, in every sense. In moments of deepest reflection, as she waited for a response to her application for the job in Hampton, Dorothy might have detected the quickening of something beyond the pragmatic hope for economic advancement, the reigniting of restless embers long quiet in the twelve years since she had come to Farmville.

Paper resolve was one thing, the messiness of real life another. She was no longer a single student with an itinerant soul but a wife and mother of four children. The job at Langley was a full-time position and required a six-day workweek at an office too far away to come home on weekends, as she had done during the summer at Camp Pickett. And yet, when the half-forgotten, hoped-for letter finally arrived, she had already made up her mind. Once Dorothy made up her mind, no one— not her husband, not her in-laws, not the principal at Moton—would be able to dissuade her from her goal.

You are hereby appointed Mathematician, Grade P-1, with pay at the rate of $2,000 per annum, for such period of time as your service may be required, but not to extend beyond the duration of the present war and for six months thereafter.

The pay was more than twice the $850 annual salary she earned teaching at Moton.

Dorothy's farewell was as straightforward and unadorned as the letter that had arrived from the NACA that fall. No party or fanfare marked her departure, just a single line in the Farmville section of the *Norfolk Journal and Guide*: "Mrs. D. J. Vaughan, instructor in mathematics at the high school for several years, has accepted a position at Langley Field, VA." Never one for the long good-bye, she lingered over her children in the house on South Main only until the bell rang at the front door. "I'll be back for Christmas," Dorothy said, with a final round of embraces. For twelve school years, every morning, she had turned left out the front door to get to work. Now the taxi turned right, spiriting her off in the opposite direction.

•   •   •

The Colored waiting room at the Greyhound bus station served as the checkpoint for an in-between world. Dorothy boarded the bus, and with each passing mile, life in Farmville faded into the distance. The job at Langley, an abstraction for half a year, moved into focus. Dorothy's previous travels—Missouri to West Virginia, Ohio to Illinois, North Carolina to Virginia—dwarfed the mere 137 miles that separated Farmville from Newport News, where she had managed to secure temporary housing using a list of rooms for rent for colored tenants. Surely she had never traveled a greater emotional distance. In the transitional space of the bus, she turned over the questions that had loitered in her mind since sending off her application six months prior. What would it be like to work with white people? Would she sit side by side with young women like the ones at the State Teachers College? Would she miss the rolling blue hills of Virginia's Piedmont, or fall in love with the great expanse of the Chesapeake Bay and the many rivers, inlets, and wetlands that embroidered the Virginia coast? How would she endure the time and distance that separated her from her children, the warmth of their embraces still fresh on her skin as the bus gained the road south?

Surrounded by grandparents and dozens of aunts and uncles and cousins, in a community where neighbors counted as family, pitching in when relatives couldn't, Dorothy's children's lives would change very little. Accustomed to their mother's long workdays and their father's extended absences, they missed Dorothy, but her departure didn't interrupt their high-spirited lives replete with family, friends, and school.

It would, however, complicate her marriage with Howard, in which time spent apart was already measured in weeks or months rather than days. Dorothy was twenty-two years old in 1932 when they married, and ready to assume the mantle of traditional family life. Dorothy, who grew up without grandparents, basked in the stability and warmth of the extended Vaughan family, but loving in-laws could provide only so much salve for a missing husband's companionship. The geographic separation between wife and husband was a proxy for the emotional distance that opened between them as the years progressed, exposing an unevenness that was perhaps present from the beginning of their relationship.

When home from the hotel circuit, Howard's longings were for the simplicities of small-town life: spending time with family and friends and working in the family's poolroom. Dorothy, on the other hand, filled every spare hour of her week with activity, from NAACP meetings to piano rehearsals at the church. Howard was satisfied with his high school diploma, but years after she chose teaching over a master's degree from Howard University, Dorothy had decided to travel to the Virginia State College for Negroes near Richmond, Virginia, once a week for a semester to take an evening extension course in education.

Dorothy, who knew the call of the open road so well, could certainly understand some of the appeal of Howard's unusual and itinerant career, and she supported it as best she could. In 1942, the entire family accompanied him to White Sulphur Springs, West Virginia, renting a house in town that was close enough for Howard to walk to his job as a bellman at the Greenbrier. Warned by their parents not to even *think* about setting foot on the hotel grounds, the Vaughan children got as close as they could to the enormous white-columned resort from the periphery, peering through the shrubbery-covered iron fence from the outside so that they might steal a glance at the German and Japanese detainees interned at a makeshift prisoner-of-war camp on the premises.

Their rented house was across the street from the home of an older Negro couple, Joshua and Joylette Coleman. Joshua and Howard shared bellman duties at the front desk of the Greenbrier. While the men worked, Dorothy and the children passed the day with Joylette, a retired schoolteacher. The Vaughan children came to love the Colemans; it was like having another set of grandparents. Dorothy, who had spent seven years of her youth in West Virginia, told stories of living in the state and listened to the Colemans' proud tales of their children's exploits, particularly those of their youngest daughter, Katherine.

Charles, Margaret, Horace, and Katherine Coleman had grown up right there in town. Twenty-four-year-old Katherine lived in Marion, Virginia, a speck of a town in the state's rural southwest. Until settling down and starting a family, Katherine had also worked as a math teacher. Like Dorothy, Katherine's intellectual gifts particularly her talent for math had skipped her ahead in school. She graduated from high school at fourteen and enrolled at West Virginia State Institute, a black college located just outside of Charleston, the state capital. By her junior year, Katherine had tackled every math course in the school's catalog and had been taken under the wing of a gifted young math professor named William Waldron Schieffelin Claytor, who created advanced math classes just for her. Claytor, who earned a PhD in math from the University of Pennsylvania in 1933, was only the third Negro in the country to obtain the credential. He had graduated from Howard University in 1929 and took a seat in the school's inaugural one-year master's degree program in mathematics—the same offer Dorothy had been unable to accept.

Whether or not Dorothy and Katherine ever realized that the brilliant Claytor was one of their shared connections—Dorothy almost never discussed her Howard admittance—Katherine's path following her graduation from the college, with a summa cum laude degree in math and French, must have felt to Dorothy like an alternate version of her own story. In 1936, the NAACP Legal Defense Fund, led by Charles Hamilton Houston, successfully argued the Supreme Court case *Murray v. Pearson*, ending graduate school admission policies that explicitly barred black students. Building on that victory, the organization scored again at the high court with 1938's *Missouri ex rel. Gaines v. Canada*, requiring states either to provide their black students with separate (but "equal") graduate and professional school programs or to allow them to integrate the white schools. Some states, like Virginia, simply refused to comply: in 1936, a black student from Richmond named Alice Jackson Houston applied to the University of Virginia to study French, but she was denied admission. The NAACP sued on her behalf, and in response, the state of Virginia set up a tuition reimbursement fund, subsidizing the graduate educations of black students in any place *but* Virginia, a policy that continued until 1950.

West Virginia, however, decided to integrate. Quietly, quickly, and without protest, three "unusually capable" Negro students began graduate studies at West Virginia University in Morgantown in the summer of 1940. The Colemans' daughter Katherine was one of them, a testament to both her academic talent and a strength of character that could stand up to the isolation and scrutiny that came along with being a black student on the front lines of desegregation. But a master's degree in math would elude Katherine just as it had Dorothy. After the summer session, Katherine decided to leave WVU's graduate program for a life as a full-time wife and mother, the call of domestic life winning out over career ambition.

Katherine's parents loved their son-in-law, Jimmy, a chemistry teacher whom Katherine met at her first teaching assignment, and they doted on their three granddaughters. Her choice to prioritize family life did nothing to dampen her parents' pride in her academic achievements. Did she, like Dorothy, ever wonder about where the opportunity might have taken her? Did she imagine what her talent might look like if it were pushed to the limit? Katherine had made her choice only two years earlier. Dorothy's first big chance was now fifteen years in the past, long enough ago to assume that the die of her life had been irrevocably cast.

And yet at the end of November 1943, at thirty-two years old, a second chance—one that might finally unleash her professional potential—found Dorothy Vaughan. It was disguised as a temporary furlough from her life as a teacher, a stint expected to end and deposit her back in the familiarity of Farmville when her country's long and bloody conflict was over. The Colemans' youngest daughter would eventually find the same second chance years in the future, following Dorothy Vaughan down the road to Newport News, turning the happenstance of a meeting during the Greenbrier summer into something that looked a lot more like destiny.

Out the window of the Greyhound bus, the gentle hills of the Piedmont flattened and broadened and the state capital came and went, and as the coastal plain of the Tidewater region advanced toward Dorothy at forty miles per hour, one of the country's busiest war boomtowns opened its arms to receive its newest resident.

CHAPTER FOUR

The Double V

Dorothy Vaughan entered the Greyhound bus in one America and disembarked in another, no less anxious, hopeful, and excited than if she were an immigrant arriving from foreign shores. The cluster of cities and hamlets around the harbor of Hampton Roads—Newport News and Hampton to the north, Portsmouth, Norfolk, and Virginia Beach to the south—boiled over with in-migrants. The region's day as a rustic land had retreated against the rolling tide of newcomers. From the forests and fisheries and farmlands of an Arcadian state dawned a powerful military capital, a nerve center that had welcomed residents by the hundreds of thousands since the start of the conflict. Now, the chief business of the people of Hampton Roads was the war.

Whether approached by land or by sea, Newport News, with its vast complex of coal piers and scaffolding, cranes and smoke-belching stacks, rails and elevators and berths laid out on the James River, gave a sense of the great power concentrated in America's military, the scope of a manufacturing and production machine of nearly inconceivable proportions, the consummation of a military-industrial empire unparalleled in the history of humankind. Stevedores and riggers by the hundreds strained against winches and loaded crates of rations and ammunition into the holds of the warships snugged into their berths.

Lines of jeeps drove onto the ships, creating traffic jams on the piers greater than any that had been seen on land. Soldiers forced teams of mules up gangways, K9 dogs boarded vessels with their faithful two-legged companions. Allied troops staged at Camp Patrick Henry, five miles up the military highway, then were delivered by train to the pier. The American mosaic was on full display, youngsters barely over the threshold of adolescence and men in the sinewy prime of manhood, fresh from the nation's cities, small towns, and countrysides, pooling in the war towns like summer rain. Negro regiments piled in from around the country. One detachment was composed entirely of Japanese Americans. Enlistees from Allied countries, like Chinese medical officers and the first Caribbean Regiment, presented themselves to the port's commanding officers before shipping out. Companies of the Women's Army Corps (WACs) stood ramrod straight and saluted. The port band sent soldiers off with "Boogie Woogie Bugle Boy," "Carolina in My Mind," "La Marseillaise"—the melodies of a hundred different hearts and hometowns.

In the boomtown, much of the work belonged to the women. The sight of coverall-clad women working at filling stations throughout the area became so common they no longer turned heads. Women shined shoes, worked at the shipyard, and staffed the offices at military installations. With men off to the front, womanpower picked up the slack, and local businesses went to extraordinary lengths to recruit and retain female employees. The War Department hired women to pose as mannequins and stand in the windows of Norfolk's Smith & Welton department store, their task to entice other women to apply for war jobs.

Between 1940 and 1942, the region's civilian population exploded from 393,000 to 576,000, and that was before accounting for the tenfold increase in military personnel, from 15,000 to more than 150,000. The war operated around the clock—three eight-hour shifts—and businesses sprinted to keep pace. Local commerce was robust—too robust in some cases: a sign reading PLEASE WASH AT HOME awaited customers of a Norfolk Laundromat enjoying too much of a good thing. The Norva Theatre in Norfolk showed movies from 11:00 a.m. to midnight, packing the house with films such as *This Is the Army* and *Casablanca*.

The flickering images offered escapism and a muscular dose of patriotism. Newsreels before and after the feature crowed about American exploits on the battlefield. Walt Disney even had an entry with an animated featured entitled *Victory Through Air Power*, extolling the virtues of the flying machine as a weapon of war. Banks, flush with cash, stayed open late to cash checks for workers. Water systems, electrical plants, school systems, and hospitals all struggled to keep pace with the growing population. Newcomers stood three deep in line for hotels, day after day. Landlords doubled their rents and still enjoyed a waiting list.

Nothing, however, quite captured the size, scope, and economic impact of the war on the Hampton Roads area like the federally funded housing development in the East End of Newport News, built to alleviate the critical shortage of homes for war workers. Migrants queued up to rent one of the 5,200 prefabricated demountable homes, 1,200 in Newsome Park, designated for blacks, and 4,000 in physically identical Copeland Park, designed for whites. From Forty-First Street to Fifty-Sixth Street, from Madison Avenue to Chestnut Avenue, the world's largest defense housing project—two smaller, separate cities within the city—took the edge off the critical housing shortage on the Virginia Peninsula.

Dorothy Vaughan arrived in Newport News on a Thursday and started work at the Langley Memorial Aeronautical Laboratory the following Monday. The personnel department maintained a file of available housing for new employees, carefully segmented by race to "establish congenial connections" and "avoid embarrassment." Five dollars a week got Dorothy a place to lay her head, two meals a day, and the kind attentions of Frederick and Annie Lucy, a black couple in their sixties. The Lucys owned a grocery store and opened their spacious home, which was located on the periphery of the Newsome Park development, to boarders. A larger version of what Dorothy had left behind, the East End was populated by stable Negro families in well-maintained homes, thriving local businesses, and a growing middle class, many of them shipyard workers whose tenure predated the boom. On the corner of the Lucys' block, a pharmacist had purchased a lot with plans to open the city's first Negro pharmacy. There was even a brand-new hospital nearby: Whittaker Memorial opened earlier in 1943, organized by black doctors and constructed by black architects.

With husband and children now far away, her living space shrunk from a spacious house to a single room, her suitcase now her closet, Dorothy's daily existence was reduced to its simplest elements. The few days of lead time were just enough to scope out the bare essentials of her new life: the location of the nearest AME church, mealtimes at the Lucy home, and transportation to work.

City buses and trolleys circulated from morning till night, swelling with riders before the orange and pink of dawn, as employees punching out from the graveyard shift met early birds just starting their day. Nowhere was the war strain more evident than in the intimate crowds of strangers who pushed up against one another in the vehicles making their rounds. Managing the multitudes in such a limited space would have been difficult under the best of circumstances, but the convoluted Jim Crow transportation laws turned the commute into a gauntlet for all riders. Whites entered and exited from the front of the bus and sat in the white section in the front. Blacks were supposed to enter and exit from a rear door and find space in back, behind the Colored line; they were also supposed to yield seats to white patrons if the white section was full. A shortage of conductors at the rear door meant that most of the time, blacks actually entered through the front door and had to push through a line of white patrons in order to get to the black section. They then jostled back through to the aisle to the front to leave the bus. And if white passengers on one of the few two-man buses found themselves at the back of the bus, they too had to push through to the front, as the law prohibited whites from using the back door. If the segregation laws were designed to reduce friction by keeping the races apart, in practice they had the opposite effect.

Overcrowded buses; a six-day workweek; constant noise and construction; shortages of sugar, coffee, butter, and meat; long lines for everything from the lunch counter to the gas station...the pressures of daily life in the boomtowns across the country pushed already touchy racial relations to the breaking point. So far, Hampton Roads had avoided the strife that had befallen Detroit, Mobile, and Los Angeles, where tensions between whites and blacks (and in Los Angeles, between Mexican, Negro, and Filipino zoot-suited youths and the white servicemen who attacked them) boiled over into violent confrontations.

Whereas white residents of the boomtowns might have seen these conflicts as caused by the war, Negroes, long conditioned to racial enmity in close quarters, were weary of the same old battles. Blacks caught sitting in white sections of buses or trolleys, no matter how crowded, were subject to fines. More than a few violators were dragged off city buses, some beaten by police. Members of a ladies' club called Les Femmes wrote a letter to the bus company complaining of the derogatory treatment their drivers routinely directed at Negro women. A bus driver on a route between Newport News and Hampton denied entry to Negro men in military uniform. Across the country, some equated the uniformed black soldiers with people who had stepped beyond their place, provoking slights and even violence against them.

Negro resistance to this injustice had been a constant ever since the first ship carried enslaved Africans to Old Point Comfort on Hampton's shores in 1609. The war, however, and the rhetoric that accompanied it created an urgency in the black community to call in the long overdue debt their country owed them. "Men of every creed and every race, wherever they lived in the world" were entitled to "Four Freedoms": freedom of speech, freedom of worship, freedom from want, and freedom from fear, Roosevelt said, addressing the American people in his 1941 State of the Union address. He committed the United States to vanquishing the dictators who would deprive others of their freedom. Negroes joined their countrymen in recoiling at the horrors Germany visited upon its Jewish citizens by restricting the type of jobs they were allowed to hold and the businesses they could start, imprisoning them wantonly and depriving them of due process and all citizenship rights, subjecting them to state-sanctioned humiliation and violence, segregating them into ghettos, and ultimately working them to death in slave camps and marking them for extermination. How could an American Negro observe the annihilation happening in Europe without identifying it with their own four-century struggle against deprivation, disenfranchisement, slavery, and violence?

Executive Order 8802 and the establishment of the Fair Employment Practices Committee brought about an upswell of optimism, with many in the black community hopeful that the gates of opportunity, finally opening, would never close again. But nearly three decades earlier, World War I had also been heralded as the event that would break the back of race prejudice. "With thousands of your sons in the camps and in France, out of this conflict you must expect nothing less than the enjoyment of full citizenship rights—the same as are enjoyed by every other citizen," President Woodrow Wilson, a native Virginian, vowed to American blacks during the previous conflict. Even then, Negroes were ready to redeem their lives for their long overdue inheritance. But the military forbade them from serving with whites, deeming them mentally deficient for the rigors of combat. Most were attached to labor battalions, as cooks and stevedores, laborers and gravediggers. The few who clawed their way into the ranks of officers still encountered filthy toilets, secondhand uniforms, segregated showers, and disrespect from white soldiers. And a man who survived the dangers of the battlefield courted danger by walking the streets of his hometown in uniform.

Charles Hamilton Houston's unyielding opposition to America's institutionalized discrimination came in part from his experiences as a young soldier in France during World War I. The man who would become the NAACP's top lawyer and the other colored soldiers in his regiment suffered endless abuse at the hands of white officers. Finally back in the United States, Houston and a friend, still in uniform, were returning home on a train when a white man refused to sit next to them in the dining car. "I felt damned glad I had not lost my life fighting for my country," he remembered in a 1942 column published in the *Pittsburgh Courier*.

After the Civil War and during the Reconstruction era, the federal government had opened jobs to blacks, providing social mobility particularly for those from educated backgrounds. Civil service reform in the late nineteenth century reduced patronage and corruption and introduced a merit system that allowed blacks to get a foot in the door. During Woodrow Wilson's presidency, however, the iron curtain of segregation fell on federal employment. A 1915 rule requiring a photo with every application made race a silent consideration for the final decision. From agencies as diverse as the Bureau of Engraving, the US Post Office, and the Department of the Navy, Wilson officials conducted a rout, purging the rolls of high-ranking black officials. Those who remained were banished to segregated areas or hidden behind curtains so that white civil servants and visitors to the offices wouldn't have to see them.

The intransigence of the forces opposed to the Negro's drive for equality was made almost unbearably plain in a 1943 comment by Mark Etheridge, editor of the *Louisville Courier-Journal*, who had served as the first head of Roosevelt's Fair Employment Practices Committee. "There is no power in the world—not even in all the mechanized armies of the earth, Allied and Axis—which would now force the Southern white people to the abandonment of the principle of social segregation," said Etheridge, a white liberal often vilified for his support of Negro advancement. The system that kept the black race at the bottom of American society was so deeply rooted in the nation's history that it was impervious to the country's ideals of equality. Restaurants that refused to serve Dorothy Vaughan had no problem waiting on Germans from the prisoner-of-war camp housed in a detention facility under the James River Bridge in Newport News. The contradiction ripped Negroes asunder, individually and as a people, their American identities in an all-out, permanent war with their black souls, the agony of the double consciousness given voice by W. E. B. Du Bois in his illuminating book *The Souls of Black Folk*.

The most outspoken members of the community refused to internalize the contradiction, openly equating the foreign racists America was moved to destroy with the American racists it chose to abide. "Every type of brutality perpetrated by the Germans, in the name of race, is visited upon the Negro in our southland as regularly as he receives his daily bread," said Vernon Johns, the husband of Dorothy Vaughan's former colleague Altona Trent Johns. The "brilliant scholar-preacher" of Farmville had gained national renown for his eloquent sermons and maverick views on racial progress. His ideas were radical for the time. However, his no-compromise policy on racial slights of any sort would have a direct and indirect influence the civil rights actions of the 1950s and 1960s.

Black newspapers—unabashedly partisan on issues pertaining to the Negro—refused to censor themselves, despite the federal government's threat to level sedition charges against them. "Help us to get some of the blessings of democracy here at home first before you jump on the 'free other peoples' bandwagon and tell us to go forth and die in a foreign land," said P. B. Young, the owner of the *Norfolk Journal and Guide*, in a 1942 editorial. As with all matters that pertained to the Negro's safety, education, economic mobility, political power, and humanity, the black press put their readers' mixed feelings about the war on full display.

James Thompson, a twenty-six-year-old cafeteria worker, eloquently articulated the Negro dilemma in a letter he wrote to the *Pittsburgh Courier*: "Being an American of dark complexion," wrote Thompson, "these questions flash through my mind: 'Should I sacrifice my life to live half American?' . . . 'Is the kind of America I know worth defending?' . . . 'Will colored Americans suffer still the indignities that have been heaped upon them in the past?' These and other questions need answering; I want to know, and I believe every colored American, who is thinking, wants to know."

What are we fighting for? they asked themselves and each other.

The question echoed off the vaulted ceilings of the auditorium at Hampton Institute's Ogden Hall. It resounded in the sanctuaries of First Baptist and Queen Street Baptist and Bethel AME and thousands of black churches around the country. It hovered in the air at the King Street United Service Organization (USO) Club, one of many centers designed to keep home-front morale high; even the USO was segregated, with separate clubs for Negroes, whites, and Jews. It dominated the headlines of the *Pittsburgh Courier*, the *Norfolk Journal and Guide*, the *Baltimore Afro-American*, the *Chicago Defender*, and every other Negro newspaper in the country. The black community posed the question in private and in public, and with every possible inflection: rhetorically, angrily, incredulously, hopefully. What did this war mean for "America's tenth man," the one in ten citizens who were part of the country's largest minority group?

It wasn't northern agitators who pushed Negroes to question their country, as so many southern whites wanted to believe. It was their own pride, their patriotism, their deep and abiding belief in the possibility of democracy that inspired the Negro people. And why not? Who knew American democracy more intimately than the Negro people? They knew democracy's every virtue, vice, and shortcoming, its voice and contour, by its profound and persistent absence in their lives. The failure to secure the blessings of democracy was the feature that most defined their existence in America. Every Sunday they made their way to their sanctuaries and fervently prayed to the Lord to send them a sign that democracy would come to them.

When American democracy beckoned them again, after the attack on Pearl Harbor, they closed ranks, as they had done in the Revolutionary War, the Civil War, the Spanish-American War, World War I, and every other American war; they geared up to fight, for their country's future and for their own. The black churches, the black sororities and fraternities, the Urban League, the National Council of Negro Women, Les Femmes Sans Souci, the Bachelor-Benedicts, black colleges across the country—they moved with an organization that shadowed the government's. The Negro press was a signal corps, communicating between leaders and the ground troops, giving the watchword so that the Negro community moved forward in sync with America, but more importantly, as a unified whole. Every action carried the hope for the ultimate victory.

From the fissure of their ever-present double consciousness sprang the idea of the double victory, articulated by James Thompson in his letter to the *Pittsburgh Courier*: "Let colored Americans adopt the double VV for a double victory; the first V for victory over our enemies from without, the second V for victory over our enemies within. For surely those who perpetrate these ugly prejudices here are seeking to destroy our democratic form of government just as surely as the Axis forces."

On the first day of December 1943, as the leaders of the United States, Great Britain, and Russia concluded a conference in Tehran in which they planned a summer 1944 invasion of France—an operation that would be known to history as D-Day—Dorothy Vaughan stepped behind the Colored line on the Citizens Rapid Transit bus and headed to her first day of work at the Langley Memorial Aeronautical Laboratory.

CHAPTER FIVE

Manifest Destiny

On her first day at Langley, Dorothy Vaughan spent the morning in the personnel department filling out the requisite paperwork. Holding up her right hand, she swore the US Civil Service oath of office, confirming her status as an employee of the National Advisory Committee for Aeronautics. But it was her employee badge—a blue metal circle dominated by an image of her face, with the winged NACA logo on either side—that sealed her status as a member of the club, the bearer of a token that allowed her free access to the laboratory's facilities. Entering the waiting Langley shuttle bus, Dorothy Vaughan headed to her final destination in the laboratory's West Area.

"If the Placement Officer shall see fit to assign thee to a far-off land of desolation, a land of marshes and mosquitoes without number known as the West Area, curse him not. But equip thyself with hip-boots, take heed that thy hospitalization is paid up and go forth on thy safari into the wilderness and be not bitter over thy sad fate," joked a contributor to the weekly employee newsletter, *Air Scoop*.

Since its establishment in 1917, the laboratory's operations had been concentrated on the campus of the Langley Field military base on the bank of Hampton's Back River. Beginning in the Administration Building, with a single wind tunnel, the lab grew until space limits pushed it to expand to the west onto several large properties tracing their provenance to colonial-era plantations. Some Hamptonites still recalled how the strange folks at the laboratory saved the town from the economic despair of Prohibition. With a disproportionate number of Hampton citizens earning a living from the liquor industry in the early days of the twentieth century, the alcohol drought that was rolling across the country was potentially devastating. The city's clerk of courts, Harry Holt, working with a cabal including oyster magnate Frank Darling, whose company, J. S. Darling and Son, was the world's third largest oyster packer, endeavored to clandestinely purchase parcels that were once the homesteads of wealthy Virginians, including George Wythe. Holt consolidated the parcels and sold them to the federal government for the flying field and laboratory. "The future of this favored section of Virginia is made," crowed the local newspaper. It was the biggest thing to happen to the area since Collis Huntington set up his shipyard in Newport News. Locals were so happy about the "life-giving energy" of federal money that they didn't even begrudge Holt and his business cronies the tidy profit they made on their real estate speculation.

Construction of the West Area began in earnest in 1939. Now, as Dorothy and the other passengers in the shuttle bus came to the end of the forested back road that connected the two sides of the campus, the view opened onto a bizarre landscape consisting of finished two-story brick buildings and cleared construction sites with half-complete structures reaching up out of what was still mostly a thicket of woods and fields. Towering behind one building was a gigantic three-story-high ribbed-metal pipe, like a caterpillar loosed from the mind of H. G. Wells. This racetrack of air called the Sixteen-Foot High-Speed Tunnel was completed just two days before the attack on Pearl Harbor and formed a closed rectangular circuit that stretched three hundred feet wide and one hundred feet deep. Adding to the futuristic aspect of the landscape was the fact that all the buildings on the West Side— indeed, all the laboratory's buildings and everything on the air base as well—had been painted dark green in 1942 to camouflage them against a possible attack by Axis forces.

The shuttle bus made the West Side rounds, stopping to deposit Dorothy at the front door of an outpost called the Warehouse Building. There was nothing to distinguish the building or its offices from any of the other unremarkable spaces on the laboratory's register: same narrow windows with a view of the fevered construction taking place outside, same office-bright ceiling lights, same government-issue desks arranged classroom style. Even before she walked through the door that would be her workaday home for the duration, she could hear the music of the calculating machines inside the room: a click every time its minder hit a key to enter a number, a drumbeat in response to an operations key, a full drumroll as the machine ran through a complex calculation; the cumulative effect sounded like the practice room of a military band's percussion unit. The arrangement played in all the rooms where women were engaged in aeronautical research at its most granular level, from the central computing pool over on the East Side to the smaller groups of computers attached to specific wind tunnels or engineering groups. The only difference between the other rooms at Langley and the one that Dorothy walked into was that the women sitting at the desks, plying the machines for answers to the question *what makes things fly*, were black.

The white women from the State Teachers College across from Dorothy's house in Farmville, and their sisters from schools like Sweetbriar and Hollins and the New Jersey College for Women, performed together in the East Area computing pool. In the West Area computing office where Dorothy was beginning work, the members of the calculating machine symphony hailed from the Virginia State College for Negroes, and Arkansas AM&N, and Hampton Institute. This room, set up to accommodate about twenty workers, was nearly full. Miriam Mann, Pearl Bassette, Yvette Brown, Thelma Stiles, and Minnie McGraw filled the first five seats at the end of May. Over the following six months, more graduates of Hampton Institute's Engineering for Women training class joined the group, as well as women from farther afield, like

Lessie Hunter, a graduate of Prairie View University in Texas. Many, like Dorothy, brought years of teaching experience to the position.

Dorothy took a seat as the women greeted her over the din of the calculating machines; she knew without needing to ask that they were all part of the same confederation of black colleges, alumni associations, civic organizations, and churches. Many of them belonged to Greek letter organizations like Delta Sigma Theta or Alpha Kappa Alpha, which Dorothy had joined at Wilberforce. By securing jobs in Langley's West Computing section, they now had pledged one of the world's most exclusive sororities. In 1940, just 2 percent of all black women earned college degrees, and 60 percent of those women became teachers, mostly in public elementary and high schools. Exactly zero percent of those 1940 college graduates became engineers. And yet, in an era when just 10 percent of white women and not even a full third of white men had earned college degrees, the West Computers had found jobs and each other at the "single best and biggest aeronautical research complex in the world."

At the front of the room, like teachers in a classroom, sat two former East Area Computers: Margery Hannah, West Computing's section head, and her assistant, Blanche Sponsler. Tall and lanky, with enormous eyes and even bigger glasses, Margery Hannah started working at the lab in 1939 after graduating from Idaho State University, not long after the East Area Computing pool outgrew the office it shared with physicist Pearl Young. Young, hired in 1922, and for the better part of two decades the laboratory's only female professional, now served as the laboratory's technical editor (the "English critic," as she was usually called) and managed a small, mostly female staff responsible for setting the standards for the NACA's research reports. Virginia Tucker, who had ascended to the position of head computer, ran Langley's entire computing operation of over two hundred women, and supervised Margery Hannah and the other section heads. The work that came to a particular section usually flowed down from the top of the pyramid: engineers came to Virginia Tucker with computing assignments; she parceled out the tasks to her section heads, who then divided up the work among the girls in their sections. Over time, engineers might bring their computing directly to the section head, or even to a particular girl whose work they liked.

With labor shortages affecting the laboratory's ability to execute time-sensitive drag cleanup and other tests designed to make military aircraft as powerful, safe, and efficient as possible, the West Computers added much-needed minds to the agency's escalating research effort. The NACA planned to double the size of Langley's West Area in the next three years. Mother Langley had even given birth to two new laboratories: the Ames Aeronautical Laboratory in Moffett Field, California, in 1939, and the Aircraft Engine Research Laboratory in Cleveland, Ohio, in 1940. Both laboratories siphoned off Langley employees, including computers, for their startup staffs. The agency scrambled to keep up with the production miracle that was the American aircraft industry, which had gone from the country's forty-third largest industry in 1938 to the *world's* number one by 1943.

For most of its existence a small and contained operation, the NACA's flagship laboratory was now a many-layered bureaucracy flush with new faces. As engineering groups grew in number and complexity, an employee's daily routine was pegged less to the revolutions of the laboratory as a whole and more to the ebb and flow of their individual work groups. Employees sat elbow to elbow with the same people during their morning coffee, ate lunch in their designated time slot in the cafeteria as a group, and left together to catch the evening shuttle bus. *Air Scoop* published everything from recaps of presentations by aeronautical notables to the scores from the intramural softball league and the dance schedule for the Noble Order of the Green Cow, the club for the laboratory's fashionable white social set. The weekly dispatch kept employees abreast of the constant activity and fostered morale, but in a breathless year in which the laboratory staff would come close to doubling, it wasn't easy for the employees themselves to absorb the full impact of the organization's unusual mission or the unusual assemblage of people carrying it out.

But just one month before Dorothy's trip from Farmville, *Air Scoop* covered Secretary of the Navy Frank Knox's one-day junket to the laboratory. Fifteen hundred employees filed into the Structures Research

Laboratory, a cavernous facility located across a dusty clearing from the Warehouse Building, to hear Knox's address. He congratulated the NACA for leading all federal agencies in employee purchases of war bonds—larger versions of the war stamps on sale at the Moton school—and lauded them for the research that turned an unreliable prototype of a dive bomber into the "slow but deadly" SBD Dauntless, a decisive force in the navy's June 1942 victory at the Battle of Midway.

"You men and women working here far from the sound of drums and guns, working in your civilian capacity in accordance with your highly specialized skills, are winning your part of this war: the battle of research," said Knox. "This war is being fought in the laboratories as well as on the battlefields."

The employees spread out from one side of the room to the other, from foreground to background, a mass occupying the enormous space like gas filling a hot air balloon. Knox, a dot at the far end of the room, stood at a podium in front of a giant American flag. White men dominated the crowd from front to back, the majority in some permutation of shirtsleeves and ties or jackets and sweaters, a good number in the coveralls of mechanics and laborers. A cluster of grandees in tweeds and armbands identifying them as minders to the secretary and his entourage stood off to the side in the front. Whiz kids of the day—John D. Bird, Francis Rogallo, John Becker, their names already circulated as being among the top in the discipline—smiled from a few rows back. Clustered in the left corner of the room stood twenty or so black men, all wearing work coats and dungarees, a few sharpening their outfits with newsboy caps or brimmed hats. White women were sprinkled throughout the crowd, many in the front row, their knee-length skirts sensibly accessorized with the practical footwear that could stand up to treks across the Langley campus. Flanking John Becker were more female faces—brown faces, peering out from the middle distance. Thelma Stiles smiled, Pearl Bassette's glasses caught the light of the flash. Tiny Miriam Mann's head was barely visible over the shoulders of the crowd. Who would have thought that such a mélange of black and white, male and female, blue-collar and white-collar workers, those who worked with their hands and those who worked with numbers, was actually possible? And who would guess that the southern city of Hampton, Virginia, was the place to find it?

After the presentation, the women of West Computing walked over to the cafeteria. Employees who never saw one another, who worked in different groups or buildings, might run into one another in the cafeteria, catch a glimpse of Henry Reid or the NACA's phlegmatic secretary, John Victory, in town for a visit, or maybe get an earful of salty language from John Stack, who oversaw the wind tunnels involved in high-speed research. Thirty minutes and back to work. Just enough time for a hot lunch and a little conversation.

Most groups sat together out of habit. For the West Computers, it was by mandate. A white cardboard sign on a table in the back of the cafeteria beckoned them, its crisply stenciled black letters spelling out the lunchroom hierarchy: COLORED COMPUTERS. It was the only sign in the West Area cafeteria; no other group needed their seating proscribed in the same fashion. The janitors, the laborers, the cafeteria workers themselves did not take lunch in the main cafeteria. The women of West Computing were the only black professionals at the laboratory— not exactly excluded, but not quite included either.

In the hierarchy of racial slights, the sign wasn't unusual or out of the ordinary. It didn't presage the kind of racial violence that could spring out of nowhere, striking even the most economically secure Negroes like kerosene poured on a smoldering ember. This was the kind of garden-variety segregation that over the years blacks had learned to tolerate, if not to accept, in order to function in their daily lives. But there in the lofty environment of the laboratory, a place that had selected them for their intellectual talents, the sign seemed especially ridiculous and somehow more offensive.

They tried to ignore the sign, push it aside during their lunch hour, pretend it wasn't there. In the office, the women felt equal. But in the cafeteria, and in the bathrooms designated for colored girls, the signs were a reminder that even within the meritocracy of the US Civil Service, even after Executive Order 8802, some were more equal than others. Even the group's anodyne title was both descriptive and a little deceptive, allowing the laboratory to comply with the Fair Employment Act—West Computing was simply a functional description on the organizational chart—while simultaneously appeasing the Commonwealth of Virginia's discriminatory separate-but-equal statutes. The sign in the cafeteria was evidence that the law that paved the way for the West Computers to work at Langley was not allowed to compete with the state laws that kept them in their separate place. The front door to the laboratory was open, but many others remained closed, like Anne Wythe Hall, a dormitory for single white women working at Langley. While Dorothy walked several blocks each morning from the Lucys' house to the bus, the women at the dormitory enjoyed special bus service. There was nothing they could do about that, or the separate "Colored girls" bathroom. But that sign in the cafeteria . . .

It was Miriam Mann who finally decided it was too much to take. "There's my sign for today," she would say upon entering the cafeteria, spying the placard designating their table in the back of the room. Not even five feet tall, her feet just grazing the floor when she sat down, Miriam Mann had a personality as outsized as she was tiny.

The West Computers watched their colleague remove the sign and banish it to the recesses of her purse, her small act of defiance inspiring both anxiety and a sense of empowerment. The ritual played itself out with absurd regularity. The sign, placed by an unseen hand, made the unspoken rules of the cafeteria explicit. When Miriam snatched the sign, it took its leave for a few days, perhaps a week, maybe longer, before it was replaced with an identical twin, the letters of the new sign just as blankly menacing as its predecessor's.

The signs and their removal were a regular topic of conversation among the women of West Computing, who debated the prudence of the action. As the sign drama played itself out in the Langley cafeteria, an incident that would have national repercussions took place in Gloucester County, just twenty miles away. Irene Morgan worked at the Baltimore-based aircraft manufacturer Glenn L. Martin Company, assigned to the production line of the B-26 Marauder. In the summer of 1944 she came home to Virginia on the Greyhound bus to visit her mother, but was arrested on the return trip to Baltimore for refusing to move to the Colored section. The NAACP Legal Defense Fund took the case and planned to use it to challenge segregation rules on interstate transportation. In 1946, the Supreme Court, in *Morgan v. Virginia*, held that segregation on interstate buses was illegal. But what hope had the West Computers of making a federal case out of something so banal as a cafeteria sign? More likely, whoever kept the table stocked with signs would just decide that it was time to get rid of the troublemakers. "They are going to *fire* you over that sign, Miriam," her husband, William, told her at night over dinner. Negro life in America was a never-ending series of negotiations: when to fight and when to concede. This, Miriam had decided, was one to fight. "Then they're just going to have to do it," she would retort.

The Manns lived on Hampton Institute's campus. Though the student body was predominantly black, the school's president and much of the faculty were white. Malcolm MacLean, a former administrator from the University of Minnesota, had taken the helm of the school in 1940, and was determined that the school's fully committed participation in the war effort would be his legacy. As the aeronautical laboratory expanded west to meet the demands of the war, its twin, Langley Field, sought to grow in order to accommodate the Army Air Corps' booming operation. A Boston philanthropist had deeded to Hampton Institute a former plantation named Shellbanks Farm, which served as an agricultural laboratory for Negro and Indian students at the school. In 1941, MacLean oversaw the sale of the 770-acre property to the federal government for use by Langley Field, making it one of the largest air bases in the world.

Under MacLean's direction, the college also established a US naval training school, effectively turning the campus into an active military base. Military police manned all campus entrances, patrolling the comings and goings of everyone on the grounds. From around the country, more than a thousand black naval recruits were sent to the school to receive instruction in the repair of airplane and boat engines. The graduates then headed off to stateside service at bases like Maryland's Naval Air Station Patuxent River, ground zero for the navy's flight test activity.

And Hampton was determined to be the leader of all black colleges in providing the Engineering, Science, and Management War Training (ESMWT) programs that had graduated West Computing's first members. Men and women crowded into Hampton Institute classrooms offering instruction in everything from radio science to chemistry. At a war labor conference that Hampton Institute hosted in 1942, MacLean told attendees that the war could be "the greatest break in history for minority groups."

Many local whites considered MacLean distastefully progressive, dangerous even, with his strident calls to boost Negro participation in the war. But it was his comfort with racial mixing in social situations that really fanned the flames. In speeches, he urged white colleges to employ Negro professors. He entertained both white and black guests at the president's residence (called the Mansion House), even allowing them to smoke. He went so far as to dance with a Hampton coed at a campus mixer, scandalizing the local gentry (and scoring points with the Hampton students). He seemed to be a true believer in the need for the Negro to advance in American society, a true champion of the tenets of the Double V.

Henry Reid, the engineer in charge of the Langley laboratory, was anything but a firebrand. An understated electrical engineering graduate of Worcester Polytechnic Institute in Massachusetts, Reid served as an able ambassador for the Yankee-heavy laboratory, replying to invitations to attend local bridge openings with the same care and promptness he used in corresponding with Orville Wright. He embraced the Hampton and Newport News Kiwanis Club set that MacLean spurned. And yet, in some ways the two men were cut from the same cloth: passionate about their particular fields, pragmatic by nature, come-heres with interests and responsibilities that extended beyond the southern sensibilities and social obligations of the town in which they worked. Almost certainly at some point they found themselves in the same place at the same time, in their hurried efforts to push their respective institutions to keep up with the rhythm of the war. Neither left fingerprints on Langley's decision to hire black women mathematicians.

Keeping a public distance from the matter might have been a strategic decision on the part of both men: if the approval process took place quietly, through the "color-blind" bureaucratic gears of the US Civil Service Commission, there was less chance of derailing an advance that served both their missions. The word about the Colored Computers made the round in the community, naturally, and there were those who saw in their employment evidence that the world was coming to an end. Even among the local gentry who attended concerts and theater at Hampton Institute's grand concert auditorium, Ogden Hall, there were those who expected to be seated at the front of the hall, apart even from the school's black faculty and administrators.

Some of Langley's white employees openly defied southern conventions. Head computer Margery Hannah went out of her way to treat the West Area women as equals, and had even invited some of them to work-related social affairs at her apartment. This was nearly unheard of, and made Marge a pariah as far as some white colleagues were concerned.

One of the most brilliant engineers on the laboratory's staff took an active interest in standing up to the prejudice he saw around town. Robert "R. T." Jones, whose theory on triangular delta-shaped airplane wings would revolutionize the discipline, was walking through the streets of Hampton one evening when he came upon a group of Hampton police officers harassing a black man. The cops were on the verge of beating the man up when Jones shouted at them to stop. They left the man alone and allowed him to leave, deciding instead to take Jones into custody. He spent the night in the city hoosegow for his trouble. Another engineer, Arthur Kantrowitz, bailed Jones out the next morning.

Engineers from the northern and western states were probably of mixed minds on the issue of race mixing. While it may have been unthinkable for most to extend their social circles to include black colleagues, within the circumscribed atmosphere of the office, they were cordial, even friendly. They got to know the women by their work, requesting their favorites for projects, open to giving a smart person— black or white, male or female—the chance to work hard and get the numbers right. That pragmatic majority, the West Computers knew, were the ones who had the power to break down the barriers that existed at Langley.

Their facilities might be separate, but as far as the West Computers were concerned, they would prove themselves equal or better, having internalized the Negro theorem of needing to be twice as good to get half as far. They wore their professional clothes like armor. They wielded their work like weapons, warding off the presumption of inferiority because they were Negro or female. They corrected each other's work and policed their ranks like soldiers against tardiness, sloppy appearance, and the perception of loose morals. They warded off the the negative stereotypes that haunted Negroes like shadows, using tough love to protect both the errant individual and the group from her failings. And each time the laboratory passed the collection plate for Uncle Sam, the West Computers reached into their purses as they had when they were teachers, so that West Computing could claim 100 percent participation in the purchase of war bonds.

At some point during the war, the COLORED COMPUTERS sign disappeared into Miriam Mann's purse and never came back. The separate office remained, as did the segregated bathrooms, but in the Battle of the West Area Cafeteria, the unseen hand had been forced to concede victory to its petite but relentless adversary. Not that the West Computers were hatching plans to invade a neighboring table; they just wanted dominion over their table in the back corner. Miriam Mann's insistence on sending the humiliating sign to oblivion gave her and the other women of West Computing just a little more room for dignity and the confidence that the laboratory might belong to them as well.

Perhaps the unseen hand and its collaborators had come to the conclusion that the quiet endurance of the West Computers was a force better engaged than antagonized, for if there was one thing the war had required over the last three years, and one thing Negroes had in abundance, it was endurance. Those forecasting a swift and tidy end to the war had been numerous but wrong. The fight slogged on, requiring more people, more money, more planes and technology. Someday the war would end, but it didn't look to be happening tomorrow. The tide of the war might be turning, but there were many battles ahead to win, and victory would require perseverance.

Not everyone could take the long hours and high stakes of working at Langley, but most of the women in West Computing felt that if they didn't stand up to the pressure, they'd forfeit their opportunity, and maybe opportunity for the women who would come after them. They had more riding on the jobs at Langley than most. The relationships begun in those early days in West Computing would blossom into friendships that extended throughout the women's lifetimes and beyond, into the lives of their children. Dorothy Vaughan, Miriam Mann, and Kathryn Peddrew were becoming a band of sisters in and out of work, each day bringing them closer to each other and tethering them to the place that was transforming them as they helped to transform it.

Dorothy listened carefully as Marge Hannah took her through the ropes of the job, taking care to note the expectations with the same exacting eye she herself had applied to grading her students at Moton: *Accuracy of operations. Skill in the application of techniques and procedures. Accuracy of judgments or decisions. Dependability. Initiative.* Even if the job lasted only six months, she was going to make the most of this chance. For an ambitious young mathematical mind—or even one not so young—there wasn't a better seat in the world.

CHAPTER SIX

War Birds

Readers of black newspapers around the country followed the exploits of the Tuskegee airmen with an intensity that bordered on the obsessive. Who said a Negro couldn't fly! Colonel Benjamin O. Davis Jr. and the 332nd Fighter Group took the war to the Axis powers from thirty thousand feet. The papers sent special correspondents to shadow the pilots as they served in the skies over Europe, each dispatch from the European front producing shivers of delight. *Flyers Help Smash Nazis! Negro Pilots Sink Nazi Warship! 332nd Bags 25 Enemy Planes, Breaks Record in Weekend Victories!* No radio serial could compete with the real-life exploits of the men who were the very embodiment of the Double V.

The "Tan Yanks," as the black press dubbed the black GIs fighting overseas, loved their planes as passionately as any other American pilots. Their lives, and those of the bomber crews they escorted, depended on knowing the plane's every strength and weakness, its peccadillos and eccentricities, on coaxing it and coercing it and waltzing with it through the sky. Initially serving in Bell P-39 Airacobras, they moved on to Republic P-47 Thunderbolts, and by the summer of 1944 the 332nd was flying North American P-51 Mustangs. "The assignment of the terrific P-51 Mustang plane to all of the Negro pilots foreshadows important missions and sweeps ahead for them as the war enters its decisive stage," wrote the *Norfolk Journal and Guide*.

"It's best described as a 'pilot's airplane,'" said an American military official in a front-page article in the *Washington Post*. "It's very fast and handles beautifully at high speeds. Fliers feel that they have always known how to fly the plane after they've been in it only a few moments." With a big four-blade propeller and a Rolls-Royce Merlin engine, the Mustang sped into the sky like a champion racehorse. Once aloft, it soared for an eternity, pushing up against 400 miles per hour with the ease of a family sedan out for a Sunday drive. And it was a damn fierce contender in a dogfight. As far as the Tuskegee airmen were concerned, it was the best plane in the world.

"I will get you up in the air, let you do your job, and bring you back to earth safely," promised the Mustang, and it delivered. Exactly how it did that wasn't the pilot's concern, but making good on that pledge was now Dorothy Vaughan's full-time job.

"Laboratories at war!" shouted *Air Scoop*. The NACA sought nothing less than to crush Germany by air, destroying its production machine and interrupting the technological developments that could hand it a military advantage. Langley was one of the United States' most powerful offensive weapons—a secret weapon, or nearly secret, hidden in plain sight in a small southern town.

Certainly the Tan Yanks would have marveled to know that supporting the performance of their beloved Mustang was a group of Colored Computers. But whereas every maneuver executed by the 332nd in their red-tailed Mustangs fed the headlines, the daily work of the West Computers and the rest of the laboratory employees was sensitive, confidential, or secret. Henry Reid advised employees to stay on the lookout for spies disguised as Langley Field soldiers and warned of fifth column plants who might coax valuable research from unwitting laboratory employees. Managers upbraided a group of messenger boys overheard dishing office dirt at a local diner, and engineers caught having a loud, detailed work conversation at the Industrial USO were called on the carpet. *Air Scoop* sounded the alarm: "You tell it to someone who repeats it to someone who's overheard by someone in Axis pay, so SOMEONE you know . . . may die!" Employees learned to keep mum on the work front even at the family dinner table. But even if they wanted to share the particulars of the day's toil, finding someone outside of Langley who understood what they were talking about would have been well nigh impossible.

In the twenty-four years since the Langley laboratory had started operation, the glitterati of the aeronautical world had made pilgrimages to Hampton. Orville Wright and Charles Lindbergh served on the NACA's executive committee. Amelia Earhart nearly lost her raccoon coat to a wind tunnel's giant turbine while touring the lab. Tycoon Howard Hughes made an appearance at the lab's 1934 research conference, and Hollywood showed up at the airfield to shoot the 1938 movie *Test Pilot*, starring Clark Gable, Spencer Tracey, and Myrna Loy. The people the famous came to see—Eastman Jacobs, Max Munk, Robert Jones, Theodore Theodorsen—were the best minds in a thrilling new discipline. Even so, most locals were oblivious to how they and their colleagues spent their days; and to be frank, they found them more than a little peculiar. Their ways and accents often marked them as Californians, Europeans, Yankees, even, God forbid, "New York Jews." They donned rumpled shirts with no ties and wore sandals; some of them sported beards. Locals dubbed them "brain busters" or "NACA nuts"; the less polite called them "weirdos."

Asked about their jobs, they demurred. Around town, they confused and horrified residents by doing things like dismantling a toaster with a screwdriver at the local department store to make sure the heating coil would toast the bread just so. One employee brought a pressure gauge from the lab into a store to test the suction capabilities of a vacuum cleaner model. Local car salesmen wanted to roll over and play dead when one of the Langley fellas pulled into the lot, fearing a barrage of nonsensical and unanswerable technical questions. They drove to work with books on their steering wheels. The NACA nuts always thought they had a better way to do anything—everything—and didn't hesitate to tell the locals so. Eastman Jacobs' legendary attempt to launch a car attached to a glider plane using Hampton's tiny Chesapeake Avenue as a runway only confirmed the Hamptonians' feelings that the good Lord didn't always see fit to give book sense and common sense to the same individual.

But Langley was a conclave of the world's best aerodynamicists, the leading edge of the technology that was transforming not only the nature of war but civilian transportation and the economy. The distance between the NACA's discovery of new aerodynamic concepts and their application to pressing engineering problems was so short, and the pace of their research and development so constant, that an entry-level position at the laboratory was the best engineering graduate school program in the world. Eager front-row boys from the lecture halls of MIT and Michigan and Purdue and Virginia Tech angled for a shot at getting in the door where Dorothy now sat.

With the goal of turning lady math teachers into crack junior engineers, the laboratory sponsored a crash course in engineering physics for new computers, an advanced version of the class offered at Hampton Institute. Two days a week after work, Dorothy and the other new girls filed into a makeshift classroom at the laboratory for a full immersion in the fundamental theory of aerodynamics. They also attended a weekly two-hour laboratory session for hands-on training in one of the wind tunnels, shouldering an average of four hours of homework on top of a six-day workweek. Their teachers were the laboratory's most promising young talents, men such as Arthur Kantrowitz, who was simultaneously an NACA physicist and a Cornell PhD candidate under the supervision of atomic physicist Edward Teller.

After twelve years at the head of the classroom, the tables had turned, and for the first time since graduating from Wilberforce University, Dorothy Vaughan gave herself fully to the discipline that had most engaged her youthful mind. She had come full circle and then some, as she tried to attune her ear to the argot that flew back and forth between the inhabitants of the laboratory, all seeking to answer the fundamental question "What makes things fly?" Dorothy, like most Americans, had never flown on a plane, and in all likelihood, before landing at Langley, she had never given the question more than a passing consideration.

The first courses imparted the basics of aerodynamics. For a wing moving through the air, the slower-moving air on the bottom of the

wing exerts a greater force than the faster-moving
difference in pressure creates lift, the almost magica
the wing, and the plane (or animal) attached to it, to asce
Smooth air flowing around the wing means the plane can
the sky with minimum friction, the way the most efficient   ers
cut through the water. Turbulent flows, like the swirl and churﬁ of rap-
ids in the water, resist the plane, slowing it down and making it harder
to maneuver. One of the NACA's great contributions to aerodynamics
was a series of laminar flow airfoils, wing shapes designed to maximize
the flow of smooth air around the wing. Aircraft manufacturers could
outfit planes with wings based on a variety of NACA specifications,
like choosing kitchen appliances from a catalog for a new house. The
P-51 Mustang was the first production plane to use one of the NACA's
laminar airfoils, a factor that contributed to its superior performance.

Future generations would take the advances for granted, but in the
early days the mechanical birds yielded their secrets slowly, pressed by
disciplined experimentation, rigorous mathematics, insight, and luck.
In the heyday of the Wright brothers and the laboratory's namesake,
inventor and researcher Samuel Langley, those with a vision for a flying
machine took a "cut and try" approach: make some assumptions, build
a plane, try to fly it, and, if you didn't die in the process, implement
what you learned on your next attempt. Aeronautics' evolution from a
wobbly infancy to a strapping adolescence gave rise to the professions
of aeronautical engineer and test pilot. Daring men—and with the exception of Ann Baumgartner Carl at Ohio's Wright Field, they were
all men—the test pilots did the "damn fool's job" of flying an airplane
directly into its weak spot. Each time the pilot pushed the aircraft to
the limit, identifying how to make a good plane better and a bad plane
nonexistent, he risked his own life and the loss of a very expensive piece
of equipment.

A wind tunnel offered many of the research benefits of flight tests
but without the danger. The basics of the tool rested on a simple concept, known even to Leonardo da Vinci: air moving at a certain speed
over a stationary object was like moving the object through the air at
the same speed. At its simplest, a wind tunnel was a big box attached

., a big fan. Engineers blasted air over planes, sometimes full-sized vehicles or fractional-scale models, even disembodied wings or fuselages, closely observing how the air flowed around the object in order to extrapolate how the object would fly through the air.

Most of the work done at Langley was of the "compressed-air" persuasion, research conducted in one of the proliferating number of wind tunnels. The names of the tunnels alone—the Variable-Density Tunnel, the Free-Flight Tunnel, the Two-Foot Smoke-Flow Tunnel, the Eleven-Inch High-Speed Tunnel—challenged the uninitiated to imagine the combination of pressure, velocity, and dimension that resided therein. The Full-Scale Tunnel's thirty-by sixty-foot test section opened wide enough to swallow a full-sized plane. Though the West Area's Sixteen-Foot High-Speed Tunnel had an exoskeleton the size of a battleship, the test section—the area where engineers, sitting at a control panel, observed the air flowing over the model—was only the size of a rowboat. But in order to accelerate the air to the necessary speed, giant wooden turbines had to accelerate the blast through the entirety of the tunnel's circuit.

Of course, while moving the air over the object was similar to flying through the air, it wasn't identical, so one of the first concepts Dorothy had to master was the Reynolds number, a bit of mathematical jujitsu that measured how closely the performance of a wind tunnel came to mimicking actual flight. Mastery of the Reynolds number, and using that knowledge to build wind tunnels that successfully simulated real-world conditions, was the key to the NACA's success. Running the tunnels during the war presented yet another logistical challenge, as the local power company rationed electricity. The NACA-ites ran their giant turbines into the wee hours if necessary, engineers pressing the machines for answers to their research questions like night owls on the hunt for mice. Residents who lived near Langley complained about the sleep-disrupting roar of the tunnels. If they'd known more about the nature of the work behind the noise, and the successes being chalked up by the strange folks next door, the neighbors might have asked for a tour.

•   •   •

No organization came close to Langley in terms of the quality and range of wind tunnel research data and analysis. The laboratory also possessed the best flight research engineers, who worked closely with test pilots, sometimes as passengers in the vehicle itself, to capture data from planes in free flight. As Dorothy learned—the West Area Computers received many assignments from the lab's Flight Research Division—it was not good enough to say that a plane flew well or badly; engineers now quantified a given vehicle's performance against a nine-page checklist under the three broad categories of longitudinal stability and control (up-and-down motion), lateral stability and control (side-to-side motion), and stalling (sudden loss of lift, flight's life force). The raw data from the work of these "fresh-air" engineers also found a home on Dorothy's desk.

What total war and the American production miracle drew into sharp relief—and what Dorothy soon learned—was the fact that an airplane wasn't one machine for a single purpose: it was a terrifically complex bundle of physics that could be tweaked to serve the needs of different situations. Like Darwin's finches, the mechanical birds had begun to differentiate themselves, branching into distinct species adapted for success in particular environments. Their designations reflected their use: fighters—also called pursuit planes—were assigned letters F or P: for example, the Chance Vought F4U Corsair or the North American P-51 Mustang. The letter C identified a cargo plane like the Douglas C-47 Skytrain, built to transport military goods and troops and, eventually, commercial passengers. B was for bomber, like the mammoth and perfectly named B-29 Superfortress. And X identified an experimental plane still under development, designed for the purpose of research and testing. Planes lost their X designation—the B-29 was the direct descendant of the XB-29—once they went into production.

The same evolutionary forces prevailed to replicate a particular model's positive traits and breed out excess drag and instability. The P-51A Mustang was a good plane; the P-51B and P-51C were great planes. After several rounds of refinement in the Langley wind tunnels, the Mustang achieved its apotheosis with the P-51D. Discoveries large and small contributed to the speed, maneuverability, and safety of the machine that symbolized the power and potential of an America that was ascending to a position of unparalleled global dominance. As the war approached its peak, every single American military airplane in production was based fundamentally—and in many cases in specific detail—upon the research results and recommendations of the NACA.

Regardless of whether the engineers conducted a test in a wind tunnel or in free flight, the output was the same: torrents, scads, bundles, reams, masses, mounds, jumbles, piles, and goo-gobs of numbers. Numbers from manometers, measuring the pressures distributed along a wing. Numbers from strain gauges, measuring forces acting on various parts of the plane's structure. If something needed to be measured and the instrument didn't exist, the engineers invented it, ran the test, and sent the numbers to the computers, along with instructions for what equations to use to process the data. The only groups that didn't run numbers based on testing worked in the small Theoretical and Physical Research Division and the Stability Research Division—the "no-air" engineers. Rather than drawing conclusions based on direct observation of a plane's performance, these engineers used mathematical theorems to model what the compressed-air engineers observed in wind tunnels and what the fresh-air engineers took to the skies to understand. The no-air girls came to think of themselves as "a cut above those that did nothing but work the machines."

What Marge passed along to Dorothy and the women of West Computing was usually a small portion of a larger task, the work by necessity carved up into smaller pieces and distributed for quick, efficient, and accurate processing. By the time the work trickled down to the computer's desk, it might be just a set of equations and eye-blearing numbers disembodied from all physical significance. She might not hear another word about the work until a piece appeared in *Air Scoop* or *Aviation* or *Air Trails*. Or never. For many men, a computer was a piece of living hardware, an appliance that inhaled one set of figures and exhaled another. Once a girl finished a particular job, the calculations were whisked away into the shadowy kingdom of the engineers. "Woe unto thee if they shall make thee a computer," joked a column in *Air Scoop*. "For the Project Engineer will take credit for whatsoever thou doth that is clever and full of glory. But if he slippeth up, and maketh a wrong calculation, or pulleth a boner of any kind whatsoever, he shall lay the mistake at thy door when he is called to account and he shall say, 'What can you expect from girl computers anyway?' "

Now and again, however, when an NACA achievement was so important that the news made the popular press, as was the case with the Boeing B-29 Superfortress, everyone got to take a victory lap. Newspapers wrote about the Superfortress and its exploits with the kind of fawning adoration accorded movie stars like Cary Grant. It was one of the planes that crossed over from being the love object of flyers and aviation insiders to a broadly known symbol of US technological prowess and bravery. The XB-29 model had logged more than a hundred hours in the laboratory's Eight-Foot High-Speed Tunnel.

"There is no one in the Laboratory who should feel that he or she did not have a part in the bombing of Japan," Henry Reid said to the lab's employees. "The engineers who assisted, the mechanics and model-makers who did their share, the computers who worked up the data, the secretaries who typed and retyped the results, and the janitors and maids who kept the tunnel clean and suitable for work all made their contribution for the final bombing of Japan."

For seven months Dorothy Vaughan had apprenticed as a mathematician, growing more confident with the concepts, the numbers, and the people at Langley. Her work was making a difference in the outcome of the war. And the devastation Henry Reid described . . . she had a part in that as well. Honed to a razor's edge by the women and men at the laboratory—flying farther, faster, and with a heavier bomb load than any plane in history—B-29s dropped precision bombs over the country of Japan from high in the sky. They brought destruction at close range with incendiary bombs, and they released annihilation—and a new, modern fear—with the atomic bombs they delivered. War, technology, and social progress; it seemed that the second two always came with the first. The NACA's work—more intense and interesting than she ever would have imagined—would remain her work for the duration. And until the war ended, whenever that might be, Dorothy would be one of the NACA nuts.

themselves ceased to exale high. The Newsome first look at the pink-pap banquet space, rooms big present, Dorothy Vaand tennis courts, and a

Or, more accurately, rk Dodgers. The center's ville, she had, once or t the Negro high schools Farmville down to her, aalization had led to his ing a school break. It wanest X-rays and diabetes whole cloth, more that ernal and civil organiza- rise, as she identified thns.

from an oscillation betwpping center included a at rest in the new city.   uty shop, a beer joint, a

Finding a suitable plasn't for sale in the stores enough supply to meet n, the milkman, the ice- most of whom considere more made the rounds, of the list of the Four Frwas a nursery school for war. Aberdeen Gardens, ng six-day weeks during by blacks" on 440 acres tme Park Elementary was Institute, had recently be was her apartment, her suburban community fonad been a young teacher. hoods like Lassiter Courer heels against the grow-

Reviewing her budgem-law that she must have job, Dorothy decided thu're not going to take my neighborhood she had cnst the changes that had best option. Although origch had roots much deeper fense employees like Dore, so did her four children, Negroes from all income Park Elementary School. business owners, and manem to ease the transition, moved in alongside the drd continued his itinerant demolition had been plarhildren on a separate path and next-door Copeland lespite the extensive travel long as the war. But the md in Farmville. He made it were built on bedrock.   as too crowded, too noisy,

Newsome Park was anr him to convince himself munity in the South, whildren back home for sum- tegration. The governmee could, unwilling and unable to sever the ties with the people she loved deeply and would always consider her family. Her marriage with Howard settled into a state of limbo, never together but never completely apart either. It was a stable instability that would endure for the rest of Howard's life, which was destined to be many decades shorter than Dorothy's.

By 1945, five out of ten people in southeastern Virginia worked for Uncle Sam, directly or indirectly. The sylvan fields, forests, and shores had been mowed down, paved over, and built up with roads, bridges, hospitals, boatyards, jails, and military bases, cities in and of themselves. Housing developments sprawled for miles, a new feature of the landscape, neither urban nor rural but something in between; the names of the new asphalted places were reflections of the green spaces they replaced: Ferguson Park, Stuart Gardens, Copeland Park, Newsome Park, Aberdeen Gardens. On the peninsula was Military Highway, a modern ribbon of road whose wide, smooth lanes now connected all the you-can't-get-there-from-here points along the finger of land from Old Point Comfort at Fort Monroe to the Newport News shipyard, with stops along the way at Langley Field and Langley. All of it was the product of the war emergency. But what was a war boomtown without the war?

V-J Day came on August 15, 1945, at 7:03 p.m. Eastern War Time. Into the vacuum of waiting and anxiety flooded "joyous tumult." All the pent-up emotions of a nation weary from four years of war exploded in a paroxysm, nowhere as much as in the war communities leading the home-front effort. From Camp Patrick Henry and Naval Station Norfolk, Langley Field and Fort Monroe, soldiers and civilians streamed into the streets. Bars and USO clubs filled in a grand hurrah. Business owners locked their doors and joined the uncounted thousands of servicemen and civilians in the celebration that lasted through the night. Spontaneous parades erupted on Washington Avenue in Newport News. In Norfolk, middies held hands and formed a human chain, dancing around cars like kindergartners, madly encircling the standstill traffic. Cries of human jubilation and "indescribable noise-making devices" sounded off into the night. Makeshift confetti snowed from windows onto the celebrants in the streets below. Some exuberant revelers piled the paper into heaps and set them on fire, the bonfires further enhancing the primal joy of the outcry. The faithful filled churches, giving thanks and imploring their creator to allow this one to be the war to truly end all wars.

After the deluge, the uncertainty settled in. Three weeks after V-J Day, the *Norfolk Journal and Guide* reported layoffs of 1,500 Newport News shipyard workers and a "decrease for women workers, both white and colored." "It seems impossible to escape the conclusion that employment in the shipyards and governmental establishments in the Hampton Roads area will be drastically curtailed," commented the *Washington Post*. Returning servicemen were expected to have first claim on what jobs remained in the peacetime economy. Just as "victory" had been the watchword for the past four years, now "reconversion" came to the fore, with the United States trying to adjust its psyche and its economy to the peace. The war had been a freight train, traveling headlong at top speed. What now of the passengers inside, still moving forward with tremendous inertia? The word "reconversion" itself implied the possibility of returning to an earlier time, of a reversal even, in the changes large and small that had transformed American life.

With the war emergency fading into the past and without war production pressures, there would be no hire-at-all-costs demand for women. Two million American women of all colors received pink slips even before the final curtain fell in August. Many anticipated a happy return to domestic life. Others, fulfilled by their work, resisted the expectation that they should be reconverted back to the kitchen and the nursery. With work had come economic security, and a greater say in household affairs, which put some women on collision courses with their husbands. "Many husbands will return home to find that the helpless little wives they left behind have become grown, independent women," wrote columnist Evelyn Mansfield Swann in the *Norfolk Journal and Guide*.

With victory over the enemies from without assured, Negroes took stock of their own battlefield. Almost immediately after V-J Day, some employers returned to their white, Gentile-only employment policies. The FEPC, however feeble it might have been in reality during the war, had nonetheless become a powerful symbol of employment progress for Negroes and other ethnic minorities. With labor markets loosening, the dream that many black leaders had of establishing a permanent FEPC slipped away with the war emergency, in spite of President Truman's support.

No one was more opposed to the FEPC than Virginia's Democratic senator, Harry Byrd, who called it "the most dangerous idea ever seriously considered" and likened it to "following the Communists' lead," an explosive epithet as the United States began to view its wartime ally Russia as the new threat. Byrd, a former governor, descended from a "First Family of Virginia," one of the state's multigenerational ruling elite. Heir to a newspaper and apple-growing fortune, Byrd treated segregation as a religion and ran a powerful political machine that kept the poor of all races divided against each other and at the bottom of the economic pyramid. "The Byrd Machine is the most urbane and genteel dictatorship in America," wrote journalist John Gunther in his 1947 bestselling book *Inside USA*. Byrd's father, who had also been a powerful state politician, had helped fellow Virginian Woodrow Wilson win the White House in 1912. It seemed too early to say if the activism and the economic gains made during the war years would carry forward into the future or give way in the face of subversion by politicians like Byrd, as they had after World War I. The generals of the Negroes' war, however—leaders such as Randolph, Houston, and Mary McLeod Bethune, who served as an advisor to President Roosevelt—did not let their guard down one bit, preparing to rouse the troops for the next offensive. But Dorothy and the others who had built new lives during the war weren't waiting for leaders or politicians to take the lead. They voted with their feet, betting their new lives that the social and economic changes brought about by the four-year conflict would last.

It wasn't a risk-free wager. Dorothy committed to the lease on the apartment in Newsome Park even though Langley had not converted her wartime employee status to permanent. The future of the neighborhood itself was also uncertain. Neighbors in nearby Hilton Village, a World War I–era housing project for white, middle-class shipyard managers, were attempting to dismantle Newsome and Copeland Parks under slum clearance laws. Federal authorities planned to pry the houses off their bases and send the units to "war-devastated populations in Europe." While the government and neighbors went back and forth over Newsome Park's status—it was declared to be "not temporary in character," yet "not permanent in its current location"—the residents brimmed with postwar idealism, calling upon each other to create a "model community, not just for Newport News, but for the entire United States." And why would Newsome Park disappear? The great groaning defense machine and all the nooks and communities it had built in the last four years weren't about to disappear. Gone were the small-town rhythms and the day of the waterman, replaced by connections to the larger world and the vitality of middle-class dreams. The jobs, the housing, the relationships, the routines—so many aspects of life that had been cut out of the whole cloth of the war emergency were now so intrinsic that it was easy to believe things had always been this way. Despite the best intentions of returning to their former lives, the come-heres tarried, realizing in small sips of awareness over the course of the war years—or with great gulping realizations at the war's abrupt end—that they would not, or could not, go home again.

Dorothy's older children had mourned the loss of their small-town freedom and the space that had come with the big house in Farmville. As talented as Dorothy was as a mathematician, she might have missed her calling in the military: she ran the Newport News household with the authority of a general and the economy of a quartermaster, eventually sending the babysitter back to Farmville and offering room and board to a returning military man and his wife in exchange for keeping the children during the day.

While her children went to school, managing the transition from being well-known faces in a small town to faces in a large crowd, Dorothy began to knit together the pieces of life she had been working on since her arrival, hosting a party for nearly twenty people in the little home on Forty-Eighth Street. Some she had met at work; others came from the neighborhood or St. Paul's AME Church. She grew closer to

Miriam Mann and her family, the two women and their children becoming like one large extended family, often taking advantage of the many activities available on the Hampton Institute campus. From the moment the acclaimed contralto Marian Anderson announced a performance at the college's Ogden Hall, the two women knew they would go together. Anderson had taken the stage there many times since her earliest professional performances as a teenager. She had gone on to sing on four continents, but there was perhaps no place she was as warmly and enthusiastically welcomed as the Hampton Institute theater; many patrons there had come out for every recital. Dorothy and Miriam Mann bought tickets in advance to secure their seats. On the evening of the concert, the Vaughans dressed up and met the Manns at the theater, arriving early so that their large group could all sit together.

It was an exceptional performance. Dorothy looked over at her children, still so young but entranced by the contralto voice that seemed to each person in the audience to be singing to them, only to them. It was, she knew right then, a moment they would never forget.

CHAPTER EIGHT

Those Who Move Forward

Katherine Goble would have eventually found her way back to the classroom, but a fever hastened the process: in 1944, her husband, Jimmy, the chemistry teacher at Marion, Virginia's Negro high school, had fallen ill with undulant fever. The illness, which came from drinking unpasteurized milk, had sickened at least eight people in Smyth County that summer. Weeks, sometimes months, of sweats, fatigue, poor appetite, and pain lay in store for the unfortunate victims. There was no way Jimmy would be able to start the school year that fall, so the principal offered Jimmy's yearlong contract to Katherine instead. Despite being a full-time wife and mother for the last four years, Katherine had been careful to keep her teaching certificate current.

It would be her second time around as a teacher at the school. In 1937, newly graduated from West Virginia State Institute, eighteen-year-old Katherine applied for a position at the Marion school, which was just on the Virginia side of the border. "If you can play the piano, the job is yours," the telegram read. She bade farewell to her home state and boarded a bus in Charleston, the state capital, settling in for the three-hour ride to Marion. Upon entering Virginia, she and the other black passengers, who had been interspersed with whites throughout the bus, were ordered to move to the back. A short time later, the driver evicted the black passengers, announcing that service wouldn't con-tinue into the town's Negro area. Katherine paid a cab to take her to the house of the principal of the Marion school, where she had arranged to rent a room.

For the two years she taught in Marion, Katherine earned $50 a month, less than the $65 the state paid similarly trained white teachers in the county. In 1939, the NAACP Legal Defense Fund filed suit against the state of Virginia on behalf of a black teacher at Norfolk's Booker T. Washington High School. The black teacher and her colleagues, includ-ing the principal, made less money than the school's white janitor. The NAACP's legal eagles, led by the fund's chief counsel, Charles Hamil-ton Houston, and Houston's top deputy, a gangly, whip-smart Howard University law school grad named Thurgood Marshall, shepherded the *Alston v. Norfolk* case to the US Supreme Court, which ordered Virginia to bring Negro teachers' salaries up to the white teachers' level. It was a victory, but a year too late for Katherine: when a $110-a-month job offer came from a Morgantown, West Virginia, high school for the 1939 school year, Katherine jumped at it. Pay equalization might have been a battle in Virginia, but West Virginia got on board without a fight.

Katherine always made sure that people knew she was from *West* Virginia, not Virginia. West Virginia's hilly terrain offered cool evening breezes, whereas Virginia was sweltering and malarial. The antebellum plantation system had never taken root in West Virginia the way it had farther east and south. During the Civil War, the mountain state se-ceded from Virginia and joined the Union. This didn't make West Vir-ginia an oasis of progressive views on race—segregation kept blacks and whites separate in lodging, schools, public halls, and restaurants—but the state did offer its tiny black population just the slightest bit more breathing room. Compared to West Virginia, the state's Negro residents thought, Virginia was the *South*.

Born and raised in White Sulphur Springs, Katherine was the youngest of Joshua and Joylette Coleman's four children. "You are no better than anyone else, and no one is better than you," Joshua told his children, a philosophy he embodied to the utmost. Dressed neatly in a coat and tie whenever he was on business in town, Joshua quietly commanded admiration from both blacks and whites in tiny White Sulphur Springs; you never had to tell anybody to respect Josh Coleman.

Though educated only through the sixth grade, Katherine's father was a mathematical whiz who could tell how many board feet a tree would yield just by looking at it. As soon as their youngest daughter could talk, Joshua and Joylette realized that she'd inherited her father's winning way with people and his mind for math. Katherine counted whatever crossed her path—dishes, steps, and stars in the nighttime sky. Insatiably curious about the world, the child peppered her grammar school teachers with questions and skipped ahead from second grade to fifth. When teachers turned around from the blackboard to discover an empty desk in Katherine's place, they knew they'd find their pupil in the classroom next door, helping her older brother with his lesson. The children's school, the only one in the area for Negroes, terminated with the sixth grade. When Katherine's older sister, Margaret, graduated from the White Sulphur Springs schoolhouse, Joshua rented a house 125 miles away so that all four children, supervised by their mother, could continue their education at the laboratory school operated by West Virginia State Institute.

Income from the Coleman farm slowed to a trickle during the hardscrabble years of the Depression. Anxious for a way to support the household and cover the cost of the children's education, Joshua moved the family into town and accepted a job as a bellman at the Greenbrier, the country's most exclusive resort. (It was here, years later, that he met Dorothy Vaughan's husband, Howard.) The enormous white-columned hotel, built in the classical revival style, sprawled on a manicured estate in the middle of White Sulphur. Joseph and Rose Kennedy spent their 1914 honeymoon in room 145 of the hotel. Bing Crosby, the Duke of Windsor, Lou Gehrig, *Life* magazine publisher Henry Luce, actress Mary Pickford, a young Malcolm Forbes, the emperor of Japan, and assorted Vanderbilts, Du Ponts, and Pulitzers all converged on White Sulphur Springs throughout the 1920s, 1930s, and 1940s, where they Charlestoned, cha-chaed, and rumbaed the night away. Even as breadlines snaked through America's main streets and drought broke the backs of tens of thousands of farm families, "Old White" remained a magnet for glamorous international guests who golfed, took the waters at the resort's famed springs, and basked in its unbridled luxury.

The Greenbrier segmented its serving class carefully. Negroes worked as maids, bellmen, and kitchen help, while Italian and Eastern European immigrants attended the dining room. During summers home from Institute, the Coleman boys pulled stints as bellmen, and Katherine and her sister took jobs as personal maids to individual guests. Accommodating the every need of the visiting gentry—cleaning their rooms; washing, ironing, and setting out their clothes; anticipating their desires while appearing invisible—was a sow's ear of a job that Katherine deftly spun to silk. One demanding French countess, with a habit of holding forth for hours on the telephone to friends in Paris, began to suspect that her walls had ears. "Tu m'entends tout, n'est-ce pas?" the countess inquired, seeing the reserved Negro maid paying close attention to her every bon mot. Katherine nodded sheepishly. The countess marched Katherine down to the resort's kitchen, and for the rest of the summer, the high school student spent her lunchtime in conversation with the Greenbrier's Parisian chef. Katherine's development from solid high school French student to near-fluent speaker with a Parisian accent astonished her language teacher that fall. The following summer the hotel placed Katherine as a clerk in its lobby antiques store, making much better use of her charisma. Henry Waters Taft, a well-known antitrust lawyer and brother to President William Howard Taft, frequented the Greenbrier, and one day in the store, Katherine taught him Roman numerals.

In 1933, Katherine entered West Virginia State College as a fifteen-year-old freshman, her strong high school performance rewarded with a full academic scholarship. The college's formidable president, Dr. John W. Davis, was, like W. E. B. Du Bois and Booker T. Washington, part of the exclusive fraternity of "race men," Negro educators and public intellectuals who set the debate over the best course of progress for black America. Though not as large or as influential as schools like Hampton, Howard, or Fisk, the college nonetheless had a solid academic reputation. Davis pushed to bring the brightest lights of Negro academe to his campus. In the early 1920s, Carter G. Woodson, a historian and educator who had earned a PhD in history from Harvard seventeen years after Du Bois, served as the college's dean. James C. Evans, an MIT engineering graduate, ran the school's trade and mechanical studies program before accepting a position as a Civilian Aide in the War Department in 1942.

On staff in the math department was William Waldron Schieffelin Claytor, movie-star handsome with nut-brown skin and intense eyes fringed by long eyelashes. Just twenty-seven years old, Claytor played Rachmaninoff with finesse and a mean game of tennis. He drove a sports car and piloted his own plane, which he once famously flew so low over the house of the school's president that the machine's wheels made a racket rolling over the roof. Math majors marveled to hear Dr. Claytor, originally from Norfolk, advancing sophisticated mathematical proofs in his drawling "country" accent.

Claytor's brusque manner intimidated most of his students, who couldn't keep up as the professor furiously scribbled mathematical formulas on the chalkboard with one hand and just as quickly erased them with the other. He moved from one topic to the next, making no concession to their bewildered expressions. But Katherine, serious and bespectacled with fine curly hair, made such quick work of the course catalog that Claytor had to create advanced classes just for her.

"You would make a good research mathematician," Dr. Claytor said to his star seventeen-year-old undergraduate after her sophomore year. "*And*," he continued, "I am going to prepare you for this career."

Claytor had taken an honors math degree from Howard University in 1929 and, like Dorothy Vaughan, had received an offer to join the inaugural class of the school's math graduate program. Dean Dudley Weldon Woodard supervised Claytor's thesis and recommended that he follow his footsteps to the University of Pennsylvania's doctoral program. Claytor's dissertation topic, regarding point-set topology, delighted the Penn faculty and was acclaimed by the mathematical world as a significant advance in the field.

Brilliant and ambitious, Claytor waited in vain to be recruited to join the country's top math departments, but West Virginia State College was his only offer. "If young colored men receive scientific training, almost their only opening lies in the Negro university of the South," commented W. E. B. Du Bois in 1939. "The [white] libraries, museums, laboratories and scientific collections in the South are either completely closed to Negro investigators or are only partially opened and on humiliating terms." But as was the unfortunate case in many Negro colleges, the position at the college came with a "very heavy teaching load, scientific isolation, no scientific library, and no opportunity to go to scientific meetings."

As if trying to redeem his own professional disappointment through the achievements of one of the few students whose ability matched his impossibly high standards, Claytor maintained an unshakable belief that Katherine could meet with a successful future in mathematical research, all odds to the contrary. The prospects for a Negro woman in the field could be viewed only as dismal. If Dorothy Vaughan had been able to accept Howard University's offer of graduate admission, she likely would have been Claytor's only female classmate, with virtually no postgraduate career options outside of teaching, even with a master's degree in hand. In the 1930s, just over a hundred women in the United States worked as professional mathematicians. Employers openly discriminated against Irish and Jewish women with math degrees; the odds of a black woman encountering work in the field hovered near zero.

"But where will I find a job?" Katherine asked.

"That will be your problem," said her mentor.

Katherine and Jimmy Goble met while she was teaching at Marion. Jimmy was a Marion native, home on college break. They fell in love, and before she headed off to West Virginia, they got married, telling no one. West Virginia might have come to equalization, but it still held the line on barring married women from the classroom.

In the spring of 1940, at the end of a busy school day, Katherine was surprised to find Dr. Davis, the president of her alma mater, waiting outside her classroom. After exchanging pleasantries with his former student, Davis revealed the motive for his visit. As a board member of the NAACP Legal Defense Fund, Davis worked closely with Charles Houston and Thurgood Marshall in the slow, often dispiriting, and sometimes dangerous prosecution of legal cases on behalf of black plaintiffs in the South. The Norfolk teachers' case was just one of many in their master plan to dismantle the system of apartheid that existed in American schools and workplaces.

In anticipation of the day that had now come, Davis, as shrewd a political operative as he was an educator, had walked away from an offer of $4 million from the West Virginia legislature to fund a graduate studies program at West Virginia State College. Davis's gamble was that if there was no graduate program at the Negro college, all-white West Virginia University would be compelled to admit blacks to its programs under the Supreme Court's 1938 *Missouri ex rel Gaines v. Canada* decision. West Virginia's Governor Homer Holt saw the writing on the wall: the choice was to integrate or, like its neighbor to the east, dig in and contest the ruling. Rather than fight, Holt moved to integrate the state's public graduate schools, asking his friend Davis in a clandestine meeting to handpick three West Virginia State College graduates to desegregate the state university, starting in the summer of 1940.

"So I picked you," Davis said to Katherine that day outside her classroom; two men, then working as principals in other parts of West Virginia, would join her. Smart, charismatic, hardworking, and unflappable, Katherine was the perfect choice. As Katherine walked out of the door on her last day at the Morgantown high school, her principal, who was also an adjunct professor in West Virginia State's math department, presented her with a full set of math reference books to use at the university, a hedge against any "inconveniences" that might arise from her need to use the white school's library.

She enrolled in West Virginia University's 1940 summer session. Katherine's mother moved to Morgantown to room with her daughter, bolstering her strength and confidence during her first days at the white school. Katherine and the two other Negro students, both men entering the law school, chatted during registration on the first day. She never saw them again on campus and sailed off alone to the math department. Most of the white students gave Katherine a cordial welcome; some went out of their way to be friendly. The one classmate who protested her presence employed silence rather than epithet as a weapon.

Most importantly, the professors treated her fairly, and she more than met the academic standard. The greatest challenge she faced was finding a course that didn't duplicate Dr. Claytor's meticulous tutelage.

At the end of the summer session, however, Katherine and Jimmy discovered that they were expecting their first child. Being quietly married was one thing; being married and a mother was quite another. The couple knew they had to tell Joshua and Joylette about their marriage and impending parenthood. Joshua had always expected that Katherine would earn a graduate degree, but the circumstances made finishing the program impossible. Katherine's love for Jimmy and her confidence in the new path her life had taken softened her father's hard line on graduate school, and he certainly couldn't resist the thrill of the family's first grandchild. Though disappointed, neither he nor the other influential men in her life—Dr. Claytor and Dr. Davis—would ever have asked her to deny love or sacrifice a family for the promise of a career.

In the four years since leaving graduate school, Katherine had not once regretted her decision to exchange the high-profile academic opportunity for domestic life. Most days she felt like the luckiest person in the world, in love with her husband and blessed with three daughters she adored. In idle moments her thoughts turned to Dr. Claytor and the phantom career he had assiduously prepared her for. In truth, the idea of becoming a research mathematician had always been an abstraction, and with the passage of time, it was easy to believe that the job was something that existed only in the mind of her eccentric professor. But in Hampton, Virginia, Dorothy Vaughan and scores of other former schoolteachers were proving that female research mathematicians weren't just a wartime measure but a powerful force that was about to help propel American aeronautics beyond its previous limits.

CHAPTER NINE

Breaking Barriers

After the war ended, the Japanese and Italian prisoners of war interned at the Greenbrier went home, but Howard Vaughan stayed on, continuing his summer work at the grand hotel alongside Joshua Coleman. The parallel lives that he and Dorothy Vaughan now led intersected often enough in Farmville and Newport News for the couple to add two more children to their family, Michael in 1946 and Donald in 1947. The youngest Vaughan children were Newport News natives, and Newsome Park was the only town they had ever known. For them, the rambling family house in Farmville was where they went on holidays and summer vacations, not a home they had left behind.

There was never a question that Dorothy would return to work as soon as possible after the births of her last two children, as soon as they were old enough to thrive in the nexus of siblings and babysitters and boarders that provided their daily lives with care and structure. An extended stay at home to care for them simply wasn't an option. The family had always counted on her income, and now, more than ever, it was her job at Langley that provided all of them with economic stability.

The older Vaughan siblings were adjusting to the changes that had expanded their lives in some ways and contracted them in others. Newsome Park came with its own set of friends and boundaries, one of

which was the pond their neighborhood shared with adjoining Copeland Park. Leonard Vaughan and his friends had it all figured out: if
they got to the pond first, it was theirs for the day. If the white kids
showed up first, they had dibs. If they both showed up at the same time,
they shared the pond, stealing curious glances at one another and making occasional small talk as they swam and played.

An adopted extended family from the West Computing office
stepped into the void left by the aunts and uncles and cousins in Farmville. Dorothy Vaughan, Miriam Mann, and the Peddrews—Kathryn
(known as "Chubby" because of her voluptuous figure) and her sister-in-
law Marjorie, who would join the office in the late 1940s—bonded over
pressure distribution curves in the office and their children and community lives outside of it. Dorothy even had a real family connection
in the group: Matilda West, related to Howard Vaughan's sister-in-law,
had followed Dorothy from Farmville to Hampton with her husband
and two young sons during the war. In the summers, the families began
the tradition of organizing a picnic at Log Cabin Beach, a wooded resort overlooking the James River, built exclusively "for members of the
race." The women spent weeks organizing the menu, hopping on and
off the phone with one another before the big day as they prepared the
culinary delights for the outing. Seven Vaughans, five Manns, and two
sets of four Peddrews, including Chubby Peddrews's dog, caravanned
up Route 60 to the riverside retreat, enjoying a rollicking day of fun
topped off by roasting marshmallows over a fire.

It felt fresh and new, that kind of free-form entertainment, away from
the traditional, structured socializing that occurred for most blacks at
home or in church, or in the nest of interconnected social and civil organizations that absorbed the precious free time of the incipient black
middle class. Negro tourists had enjoyed sun and fun and amusement
at Hampton's Bay Shore Beach ever since a group of black businessmen,
including Hampton Institute's bookkeeper and local black seafood en-
trepreneur John Mallory Phillips, founded the resort in 1898. But Bay
Shore, separated as it was by a rope from larger, whites-only Buckroe
Beach, still reminded clients that there was Negro sand and there was

white sand. At Log Cabin, Negroes with the means to take advantage of its charms left the "colored" signs behind completely. They could occupy the entirety of the space, free from the signs constricting their physical movements and the double consciousness that throttled their souls.

Dorothy loved allowing her children to take unguarded steps into the world; having access to a broader base of experiences was one of the most compelling reasons for turning their world on its head with the move to Hampton Roads. Even with a salary of $2,000 a year—the average monthly wage for black women in the 1940s was just $96—providing for the needs of six children meant that outings like the ones at Log Cabin Beach did not come often or easily. With the shadow of the Depression always at the back of her mind, Dorothy Vaughan sewed clothes for herself and her children, clipped coupons, and wore shoes until her feet started to push through the worn soles. If she could give more to her children by sacrificing her own comforts, she did it. Many was the evening when she came home from work to make dinner, and after putting the meal on the table, walked out the door and took a walk around the block until the children were done eating. Only then would she serve herself from the leftovers. She didn't want to face the temptation of eating even one morsel herself that could nourish their growing bodies.

The prediction that the end of the war would send Hampton Roads into an economic downturn proved incorrect. The place Dorothy Vaughan now called home was on the cusp of a defense industry boom that would be measured not in years but in decades. After the war, the Norfolk Naval Base confirmed its command of the Atlantic Fleet and was appointed the headquarters of the navy's air command. Added to the local military installations and contractors were the Army Transportation School, set up at Fort Eustis in Newport News, and the US Coast Guard base in Portsmouth, with the Newport News shipyard and the naval shipyard in Portsmouth still going strong. In 1946, the army decided to make Langley Field the headquarters of its Tactical Air Command, one of the major commands of the US Army Air Corps.

One year later, the importance of the airplane to US defense was underscored when the Army Air Corps was elevated to the status of an independent branch of the military: the United States Air Force.

The defense establishment's grip on the economy of southeastern Virginia had become so strong during the war, its influence so critical to the material well-being of the local residents, that just as Hampton Roads was once the very model of the war boomtown, so too had it become a warfare state, dependent on the defense industry dollars that lapped into the region like waves on the shores of its beaches. Hampton Roads had become the embodiment of what Cold War president Dwight D. Eisenhower would a decade later dub the "military-industrial complex."

The inevitable reduction in force that visited Langley—the staff had peaked at more than three thousand employees just before V-J Day—was short-lived and shallow, mostly accomplished through the natural attrition of those who decided it was time to move on from their life at Langley. Many computers and other female employees at the laboratory exchanged the daily routine of office life for a full-time position at home. No small number of them married the men they worked with; the laboratory's success as a matchmaking service rivaled its research prowess. The employee newsletter *Air Scoop* was replete with tales of the sparkler spotted on the ring finger of this bachelorette in the personnel department or the happy ending for that starry-eyed couple who found true love over model tests in the Free-Flight Tunnel.

A steady stream of "heir mail" announcements ensued. Disability or accumulated sick leave were available to expectant mothers who wanted to return to work when their children were old enough to be left in the care of someone else during the day, though how easily this was done depended on the disposition of their managers. Many women tendered their resignations during their pregnancies and reapplied to the laboratory when they were ready to work again, hoping they would find a way to wrangle back their old jobs.

But talented computers, particularly those with years of experience, were valuable resources. The ink was barely dry on the bulletin in *Air Scoop* announcing the reduction in force before Melvin Butler released a plan to offer permanent appointments to war service employees. Some top-ranked managers went out of their way to keep the most productive women on the job by giving them the flexibility they needed to take care of their families.

In three years at Langley, Dorothy Vaughan had proven to be more than equal to the job, handing off error-free work to Marge Hannah and Blanche Sponsler and managing the constant deadlines with ease, garnering "excellent" ratings from her bosses. During the war, Dorothy and two other colleagues, Ida Bassette, a Hampton native (cousin to West Computing's Pearl Bassette), and Dorothy Hoover, originally from Little Rock, Arkansas, had been appointed shift supervisors, each managing one-third of a group that had swelled to twenty-five women. At the war's peak, when the laboratory operated on a twenty-four-hour schedule, Dorothy often worked the 3:00 p.m.-to-11:00 p.m. shift, responsible for eight computers' work calculating data sheets, reading film, and plotting numbers. Perhaps it was no surprise that Dorothy was a keeper, but it must have come as a relief when, in 1946, she was made a permanent Civil Service employee.

Nearly to a woman, the West Computers had decided they were going to hold on to their seats, whatever it took. The section had outgrown its original room in the Warehouse Building, and in 1945, the group moved into "two spacious offices" on the first floor of the West Side's newly built Aircraft Loads Division Building.

Black or white, east or west, single or married, mothers or childless, women were now a fundamental part of the aeronautical research process. Not a year after the end of the war, the familiar announcements of vacancies at the laboratory, including openings for computers, began to appear in the newsletter again. As the United States downshifted from a flat-out sprint to victory to a more measured pace of economic activity, and as the laboratory began to forget that it had ever operated without the female computers, Dorothy had time to pause and give consideration to what a long-term career as a mathematician might look like. How could she entertain the idea of returning to Farmville and giving up a job she was good at, that she enjoyed, that paid two or three times more than teaching? Working as a research mathematician at Langley was a *very, very* good black job—and it was also a *very, very* good female job. The state of the aeronautics industry was strong, and the engineers were just as interested in retaining the services of the women who did the calculations as the aircraft manufacturers had been in keeping the laundry workers who supported their factory workers on the job.

Thousands of women around the country had taken computing jobs during the war: at Langley; at the NACA's other laboratories (the Ames Research Laboratory in Moffett Field, California, which the agency had founded in 1939, and the Flight Propulsion Research facility in Cleveland, inaugurated in 1941); at the army's Jet Propulsion Laboratory, run by the California Institute of Technology; at the Bureau of Standards in Washington, DC; at the University of Pennsylvania's secret ballistic research laboratory; and at aircraft companies like Curtiss Wright, which called the women "Cadettes." A new future stretched out before them, but Dorothy Vaughan and the others found themselves at the beginning of a career, with few role models to follow to its end. Just as they had learned the techniques of aeronautical research on the job, the ambitious among them would have to figure out for themselves what it would take to advance as a woman in a profession that was built by men.

Renowned aerodynamicists Eastman Jacobs, John Stack, and John Becker had come to the laboratory as freshly minted junior engineers and were quickly allowed to design and conduct their own experiments. R. T. Jones, the engineer who had intervened on behalf of the black man with the Hampton police, had beguiled Langley managers immediately with his facile mind. Jones never finished college, so when he was hired in 1934, it was as a subprofessional scientific aide, the category that most women fell into. Despite good performance reviews, a P-1 rating was out of his reach, because the grade required a bachelor's degree. So his managers conspired to skip him ahead to a P-2, which, by the arcane rules of bureaucracy, didn't have the same requirement.

Seasoned researchers took the male upstarts under their wings, initiating them into their guild over lunchtime conversations in the cafeteria and in after-hours men-only smokers. The most promising of the acolytes were tapped to assist their managers in the operations of the laboratory's valuable tunnels and research facilities, apprenticeships that could open the door to high-profile research assignments and eventual promotion to the head of a section, branch, or division. By the end of the 1930s, R. T. Jones had been promoted to head the Stability Analysis section, an influential redoubt of the "no-air" engineers, which used theoretical math rather than wind-tunnel experiments or flight tests to understand how to improve aircraft performance.

Women, on the other hand, had to wield their intellects like a scythe, hacking away against the stubborn underbrush of low expectations. A woman who worked in the central computing pools was one step removed from the research, and the engineers' assignments sometimes lacked the context to give the computer much knowledge about the afterlife of the numbers that bedeviled her days. She might spend weeks calculating a pressure distribution without knowing what kind of plane was being tested or whether the analysis that depended on her math had resulted in significant conclusions. The work of most of the women, like that of the Friden, Marchant, or Monroe computing machines they used, was anonymous. Even a woman who had worked closely with an engineer on the content of a research report was rarely rewarded by seeing her name alongside his on the final publication. Why would the computers have the same desire for recognition that they did? many engineers figured. They were *women*, after all.

As the computers' work grew in scope and importance, however, a girl who impressed engineers with her mathematical prowess might be invited to join them working full-time for their tunnel or group. More groups meant more opportunities for women to get closer to the research and establish their bona fides. Computing pools attached to specific tunnels or branches grew larger, spawned their own supervisors, and gave the female professionals the opportunity to specialize in a particular subfield of aeronautics. A computer who could process data on the spot and understand how to interpret it was more valuable to the team than a pool computer with more general knowledge. That kind of specialization would be the key to managing the increasingly complex nature of aeronautical research in the postwar era. Freed from the wartime imperative of drag cleanup—the process of refining existing planes to eke out small improvements in their performance—the aerodynamicists turned their attentions to an enemy more difficult to defeat than the Axis forces: the speed of sound.

The development of the turbojet engine in the early 1940s meant that Langley engineers finally had a powerful enough propulsion system to make their high-speed wing concepts, like R. T. Jones's swept-back delta wings, which were angled backward like the wings of a swift—a high-flying bird—*really* fly. Langley added state-of-the-art facilities on the West Side like the Seven-by-Ten-Foot High-Speed Tunnel and the Four-by-Four-Foot Supersonic Pressure Tunnel, machines that could blast models with winds that approached or exceeded the mysterious speed of sound. The NACA empire also continued the course of its westward expansion, building up its staff and facilities at the Cleveland and Ames laboratories.

In 1947, a party of thirteen Langley employees, including two former East Computers, was sent to the Mojave Desert to establish the Dryden High-Speed Flight Research Center, a direct assault on the problems of faster-than-sound flight. The speed of sound, about 761 miles per hour at sea level in dry air at 59 degrees Fahrenheit, varied depending on temperature, altitude, and humidity. It was long thought to be a physical limit on the maximum speed of an object moving through the air. As an airplane flying at sea level in dry air approached Mach 1, or 100 percent of the local speed of sound, air molecules in front of the flying plane piled up and compressed, forming a shock wave, the same phenomenon that caused the noise associated with the crack of a bull whip or the firing of a bullet.

Some scientists speculated that if a pilot succeeded in pushing his plane through the sound barrier, either the plane or the pilot or both would disintegrate from the force of the shock waves. But on October 14, 1947, pilot Chuck Yeager, flying over the Mojave Desert in an NACA-developed experimental research plane called the Bell X-1, pierced the sound barrier for the first time in history, a fact that was corroborated by the female computers on the ground who analyzed the data that came from the instruments on Yeager's plane.

There were too few women at Muroc to warrant sending them off into a separate section. In the isolation of the desert, in close working conditions at a bare-bones facility with ramshackle dormitories, the Muroc computers stepped easily into the role of junior engineers. Upward mobility was more difficult to achieve in the larger, more bureaucratic operation at Mother Langley, with a well-developed management structure. Even there, however, a few pioneers were managing to clear a path of sorts for other women to follow. Mathematician Doris Cohen, a native New Yorker who started working at the laboratory in the late 1930s, was for many years the NACA's lone female author. Not even Pearl Young, the NACA's first female engineer and the founder of the agency's rigorous editorial review process, left behind research with her name on it.

From 1941 through 1945, Doris Cohen published nine reports documenting experiments conducted at the frontier of high-speed aeronautical research, five as the sole author, and four coauthored with R. T. Jones (whom she would eventually marry). It was the kind of prodigious output that even aspiring male engineers could only hope to replicate. Getting one's name on a research report was a necessary first step in the career of an engineer. For a woman, it was a significant and unusual achievement. It provided public acknowledgment that she had contributed to a worthy line of inquiry and inked her fingerprints onto findings that would be circulated widely among the aeronautical community. Authors of a report were identified as important members of a team; proximity to the work was everything. As more of the women in the computing pools transferred to engineering groups—and as new computers were hired into sections from their first day of work, without serving time in the pools—it gave the women the chance to move away from "working the machines" and rote plotting and to get closer to the research report, which was the laboratory's most important product.

The strongest evidence of the progress Langley's women were making in the early postwar years came when one of its most visible female professionals reached the end of the road. Over the course of twelve years, Virginia Tucker, the lab's Head Computer, had ascended from a subprofessional employee to the most powerful woman at the lab. She had done much to transform the position of computer from a proto-clerical job into one of the laboratory's most valuable assets. Her relentless recruitment efforts at the Women's College of the University of North Carolina—in 1949, the largest all-female college in America—and other all-female schools had given hundreds of educated women a shot at a mathematics career. All of the agency's computing staffs, at Langley, Cleveland, Ames, and Muroc, traced their lineage back to the first pool, and to Tucker's labor as the first female computer supervisor. Between 1942 and 1946, four hundred Langley computers received training on Tucker's watch.

The East Computing of the war years, a section with so many employees that the women had been forced to set up in hallways or closets or anywhere that could accommodate them, now fell victim to its own success. Veteran East Computers left for permanent assignments in the tunnels, and no new girls were hired into the pool to replace them. The core group, which had been housed in an office in the Nineteen-Foot Pressure Tunnel on the East Side, dwindled away. Girls now reported directly to their engineers or to the computer supervisors attached to the group. Virginia Tucker was a respected manager, but unlike Doris Cohen, she hadn't pursued the researcher's path and had no research credit to her name. She held a senior position for a woman but found herself without an obvious next step at Langley. In 1947, the laboratory disbanded East Computing, rerouting all open assignments to West Computing. Virginia Tucker, too, decided to head west. She accepted a job at the Northrop Corporation, one of many aviation companies tucked into the suburban sprawl of Los Angeles. The company hired her as an engineer.

While the East Computers had flowed out of their office and into the larger operations of the laboratory like a river, segregation kept the out-migration of West Computing to a trickle. When three West Computers made the leap in the late 1940s to Cascade Aerodynamics, a group that studied the aerodynamics of propellers, turbines, and other rotating bodies, it caused a commotion. Many white laboratory employees, particularly on the East Side, hadn't even known that an all-black computing group existed. A conservative minority saw the mathematical race mixing as the twilight of their civilization. But the West Computers' competence silenced most of the dissent; up close, it was difficult to object to good educations and mild middle-class manners, even if they came wrapped in brown skin.

It was inevitable that one of the black women would get a shot at a research job. Dorothy Hoover—West Computing's other Dorothy, one of the three shift supervisors—had earned an undergraduate degree in math from Arkansas AM&N, a black college that had been active in the World War II ESMWT training programs. She knocked off a master's degree in mathematics from Atlanta University, and then taught school in Arkansas, Georgia, and Tennessee before coming to Langley in 1943, where she was hired as a P-1 mathematician. Like Doris Cohen, Dorothy Hoover was exceptionally fluent in abstract mathematical concepts and complex equations, and Marge Hannah channeled the rigorous mathematical assignments that came to the group from R. T. Jones' Stability Analysis Section to Hoover. The engineers provided Hoover with long equations defining the relationships between wing shape and aerodynamic performance and instructed her to substitute into them other equations, formulas, and variables. Only when the series of equations had been sufficiently reduced would she start the process of inputting values and coming up with numbers using the calculating machines.

The Stability Analysis fellows were as well known for their progressive politics as for their sharp minds. Many of them were Jewish, from the North. Jones, his wife, Doris Cohen, along with Sam Katzoff and Eastman Jacobs, who were two of the laboratory's most respected analysts, and a white economics professor at Hampton Institute named Sam Rosenberg often convened at the home of Langley researcher Arthur Kantrowitz, where they spent evenings listening to classical music and discussing politics. They went to see movies at the Hampton Institute theater, comfortably mingling at the Negro school. More than most, they were open to breaking the color barrier by integrating a West Computer into their group. As a talented, possibly gifted mathematician with a similarly independent mind, Dorothy Hoover was a perfect fit for the section, and in 1946 R. T. Jones invited her to work directly for him.

With East Computing's calculating added to existing workload, Dorothy Vaughan and the West Area Computers remained in high gear. The laboratory was still hiring black women into the pool faster than they were being sent out to other positions. Women who did get assigned to another section were usually staffed on temporary duty and eventually returned, keeping the two offices full, at least for the moment.

After the war, West Computing's head, Margery Hannah, decided to accept an offer from the Full-Scale Research Division, an excellent assignment working for Sam Katzoff. Within three years, she would join the ranks of female authors, publishing a study with Katzoff attempting to measure the degree to which waves bouncing off wind tunnel walls interfered with the airflow over a model. Like sound waves in an auditorium or water slapping up against the sides of a swimming pool, the gusts in a wind tunnel ricochet off the enclosure, and test results had to account for discrepancies caused by the interference.

Marge's upward move resulted in opportunity down the line: Blanche Sponsler stepped into Marge's position as head of the group. Just two years younger than Dorothy, Blanche was a thirty-five-year-old newlywed in 1947. Originally from Pennsylvania, she bowled in Langley's Duckpin bowling league and was a faithful member of the Bridge Club. She and her sister, the wife of a soldier stationed at Fort Monroe, entered the laboratory's Duplicate Bridge tournament in 1947 and took second place. Also interested in a move west, Blanche requested a transfer to the Ames Laboratory. Her Langley supervisors wrote letters of recommendation—since coming to the lab in 1940, she had received strong reviews and steady promotions—but at the time there were no open positions, so she continued as head of the West Computing group.

Dorothy had worked with Blanche since 1943. They enjoyed a good professional relationship, and Blanche gave Dorothy strong performance ratings. Dorothy's role as a shift supervisor raised her profile with engineers. Part consultants and part teachers, computing supervisors had to be top-notch computers themselves, capable of grasping the needs of the engineer and clearly explaining the requirements to her subordinates. She fielded the computers' questions and needed a strong-enough command of the math to tutor the women through any weaknesses. Tapping the right girl for a particular job was a big part of the manager's responsibility. All the women were proficient in basic computing tasks, but knowing who was a perfectionist with the computing machines and who could churn out perfect graphs on short order was key to processing the data most efficiently. A small elite, like Dorothy Hoover, were endowed with an aptitude for complex math so strong that it exceeded the ability of many of the engineers at the lab.

Dorothy Vaughan might have eventually lobbied to follow Margery Hannah and Dorothy Hoover into a job working directly for an engineering section. As a supervisor she came into contact with engineers from a variety of groups, some of whom came to the office insisting that she personally handle their jobs. In 1949, however, an unusual and tragic turn of events would bind Dorothy to the West Area computing office for the next decade.

At the end of 1947, Blanche had left the group in Dorothy's command during a one-month illness. She returned to work, appearing none the worse for wear, but was out of work again on a leave of absence during July and August 1948. This time, too, she returned to the office, snapped back into her routine, and continued uneventfully for the next several months. But on the morning of January 26, 1949, a West Computer made an urgent call to Eldridge Derring, one of the lab's administrators. For the last few days, she told Derring, Blanche had been acting strangely. Now, Blanche was in the office "behaving irrationally," and she implored him to come to the Aircraft Loads Building to help the women deal with the situation. Derring, along with the lab's health officer, James Tingle, and Rufus House, assistant to Langley director Henry Reid, hustled over to the building, where several West Computers were anxiously waiting in the lobby.

Together, they all went into one of the West Computing offices, where Blanche was standing in the middle of the room, preparing for a 10:00 a.m. meeting. She had covered the blackboard in the office with "meaningless words and symbols" and began to conduct the meeting in what she seemed to feel was normal fashion. However, she was completely unintelligible to the people in front of her. House approached Blanche to ask about the gibberish covering the board.

"I'm trying to explain how to go from SP-1 to P-20," she told him, adding that the number of SP-1 employees in her group was "0 ±1 to three significant figures," and that there was "one P-75,000" in the section. She then said that she was trying to explain the difference between zero and infinity. ("Quite rational," commented House afterward, in a memo detailing the morning, "as some college students have had difficulty in comprehending this difference.") The rest of Blanche's diatribe declined from there. House asked her to accompany him to the East Area, hoping to take her to the psychiatrist on the air force base hospital. She refused to leave, but the men didn't force the issue, concerned that if she was provoked and became violent, it would require "at least four strong men" to subdue her. Finally, quietly, Blanche turned her back on the group. She began to weep, wiping her eyes with a handkerchief. The administrators dismissed the meeting, and the other women filed into the other office, leaving Blanche alone with the men.

In the hush-hush 1940s, such a public display of mental illness would have spelled the end of Blanche's career at Langley, even if she had been able to recover from the episode. That afternoon, Blanche Sponsler was taken away to the Tucker Sanatorium, located in Richmond, the state capital. She had been admitted to the same hospital for treatment during her 1948 hiatus, and presumably this problem was also the reason behind her absence in 1947. She languished in the Tucker Sanatorium for three months before being transferred to Eastern State Hospital in Williamsburg. This time, she was incapable of returning to her previous life.

"It appears that she will continue ill indefinitely," Eldridge Derring commented to Langley's personnel officer, Melvin Butler, two weeks later.

The women in West Computing never saw Blanche Sponsler again. A June 3, 1949, note in *Air Scoop* served as the only postscript to their former supervisor's tenure at the laboratory: "Blanche Sponsler Fitchett, Head of West Computing Section, died last Sunday after a six-month illness." The cause of her death, not revealed in the note or in her obituary in the *Daily Press*, was entered on her death certificate as "dementia praecox." Whether Blanche died as a result of treatments designed to cure an illness that would eventually become known as schizophrenia or from suicide or another cause altogether was known only to her doctors and family.

Blanche's absence left West Computing with an empty desk, but not a vacuum. It wasn't the way Dorothy would have wanted to take the next step in her career, but Blanche's tragedy pushed her up the ladder nonetheless. In April 1949, six weeks after Blanche left the office for the last time, the laboratory appointed Dorothy Vaughan acting head of West Computing.

There were limited ways for a white computer to break into management at Langley. Finding a way to move from being one of the girls to one of the Head Girls took time and persistence, pluck and luck, and there were only so many slots available: while even lower-level male managers might supervise the work of female computers, it was simply unthinkable for a man to report to a woman. Women with an eye on a management job were limited to heading a section in one of the now-decentralized computing pools or in another division with many female employees, such as personnel.

For a black woman, there was exactly one track: it began at the back of the West Area computing office and ended at the front, where Dorothy Vaughan now sat. The view from the supervisor's desk, with the rows of brown faces looking back at their new boss, wasn't that different from being at the head of the classroom at Moton: the segregation laws of the state applied just as vigorously to the roomful of highly educated college graduates as they did to the rural black students of Prince Edward County. Yet with its bright lights, government-issue desks, late-model calculating machines, and proximity to tens of millions of dollars' worth of aeronautical research tools, West Computing was a world away from Moton High School's deficient building, rundown chairs, worn-out textbooks, and general sense of powerlessness.

It would take Dorothy Vaughan two years to earn the full title of section head. The men she now worked for—Rufus House was her new supervisor—held her in limbo, waiting either until a more acceptable candidate presented herself or until they were confident she was fit to execute the job on a permanent basis. Or maybe the idea of installing the first black manager in all of the NACA's expanding national empire caused them to demur, lest they stoke the racial anxieties among members of the laboratory and in the town.

Whatever skepticism might have existed among the powers that be about Dorothy's qualifications, whatever lobbying and advocacy may have been required on Dorothy's part, the outstanding issue was resolved by a memo that circulated in January 1951. "Effective this date, Dorothy J. Vaughan, who has been acting head of the West Area Computers unit, is hereby appointed head of that unit." Dorothy must have known it. Her girls and her peers knew it. Many of the engineers knew it, and her bosses eventually came to the same conclusion. History would prove them all right: there was no one better qualified for the job than Dorothy Vaughan.

CHAPTER TEN

Home by the Sea

In April 1951, as the laboratory shuttle transported twenty-six-year-old Mary Winston Jackson from new employee processing in the personnel department over to West Computing, virtually no evidence remained of the agricultural roots of the land that had become Langley. The come-heres like Dorothy Vaughan and her band of sisters, like the phalanx of Yankees and Mountaineers and Tar Heels who had descended upon the laboratory during the war, would tell a lifetime of stories about the transformations they witnessed as Hampton Roads emerged from agrarian isolation to become a vibrant collection of cities and defense industry suburbs. But Mary Jackson remembered the prewar hamlet where Negro vacationers still made their way to Bay Shore Beach by trolley car. She grew up listening to the work songs of the black women shucking oysters at the J. S. Darling processing plant that wafted up to the pedestrians on the Queen Street Bridge above. During Mary's child-hood, elders at the black churches in the heart of downtown Hampton still told stories of sitting under a glorious oak tree across the river, on the campus of what would become Hampton Institute, and listening to Union soldiers read the Emancipation Proclamation. Those ancestors walked into the gathering as legal property and emerged as free citizens of the United States of America. No one was more of a been-here than Mary Jackson.

The Olde Hampton neighborhood where Mary grew up, in the heart of downtown, was literally built upon the foundations of the Grand Contraband Camp, founded by slaves who had decided to liberate themselves during the Civil War from the families that had stolen their labor and their lives. The refugees sought shelter as "contraband of war" in the Union stronghold at Fort Monroe, located at Old Point Comfort, on the tip of the Virginia Peninsula. The freed colored people raised central Hampton from the ashes of the "Confederate-set inferno" that consumed the city in 1862. Olde Hampton's street names—Lincoln, Grant, Union, Liberty—memorialized the hopes of a people fighting to unite their story with the epic of America. In the optimistic years after the Civil War, before the iron curtain of Jim Crow segregation descended across the southern United States, Hampton's black population earned a measure of renown for its "educated young people, ambitious and hardworking adults, its successful businessmen, and its skillful politicians."

It was no small irony that Woodrow Wilson, the president who had authorized the creation of the NACA and who received a Nobel Peace Prize for his promotion of humanitarianism through the League of Nations, was the very same one who was hell-bent on making racial segregation in the Civil Service part of his enduring legacy. Now, Mary's presence at the laboratory built on plantation land rebuked the short-sighted intolerance of her fellow Virginian. Mary's family, the Winstons, had the same deep Hampton roots as Pearl and Ida Bassette. Mary's sister Emily Winston had worked with Ophelia Taylor in the same nursery school during the war, before Taylor headed off to the Hampton Institute training program. Many of the West Computers, including Dorothy Vaughan, were members of Alpha Kappa Alpha, the sorority that Mary had pledged as an undergraduate at Hampton Institute. Mary graduated in 1938 with highest honors from Phenix High School. Phenix, located on the Hampton Institute campus, was like the upper school that Katherine Goble attended on the campus of West Virginia State. It served as the de facto public secondary school for the city's Negro students, since the city provided schooling for them only through elementary school. Mary followed the family tradition of enrolling in Hampton Institute, which had graduated her father, Frank Winston, her mother, Ella Scott Winston, and several of her ten older siblings. The school's philosophy of Negro advancement through self-help and practical and industrial training—the "Hampton Idea," closely associated with Booker T. Washington, the college's most famous graduate—mirrored the aspirations and philosophy of the surrounding black community.

Most of Hampton Institute's female students earned their degrees in home economics or nursing, but Mary Jackson had a strong analytical bent, and she pushed herself to complete not one but two rigorous majors, in mathematics and physical science. She intended to put her degree to use as a teacher, of course; there were practically as many teachers in her family as there were Hampton Institute graduates. She fulfilled her student teaching requirements at Phenix High, and after graduating in 1942, accepted a job teaching math at a Negro high school in Maryland. At the end of the school year, however, she returned to Hampton to help care for her ailing father. Nepotism laws forbade her from teaching in one of Hampton's public Negro elementary schools, since the school system already employed two of her sisters. But her excellent organizing skills, fluency with numbers, and good marks in a college typing course made her the perfect fit for the King Street USO, which in 1943 was looking for a secretary and bookkeeper.

While the women in Hampton Institute's Engineering for Women courses were preparing for their new careers as computers, Mary Jackson managed the USO's modest financial accounts and welcomed guests at the club's front door. Her daily schedule, however, usually overflowed well beyond the job's narrow duties, since the club quickly became a center for the city's black community. She helped military families and defense workers find suitable places to live, played the piano during the USO's rollicking singalongs, and coordinated a calendar of Girl Scout troop meetings and military rallies. She organized dances at the club, making sure that the Junior Hostesses and Victory Debs were on hand to entertain visiting servicemen. The people who came to the club for movie night or cigarette bingo, for tips on where to worship or get their hair cut, or just a hot cup of coffee, appreciated the energy, warmth, and can-do skill of the young woman at the front desk. If Mary Jackson didn't know how to get something done, you could bet a dollar to a USO doughnut she'd find the person who did.

Her family's motto was "sharing and caring," and even in a community of active citizens, the Winstons distinguished themselves with their tireless service, religious devotion, and humanitarianism. Mary's father, Frank Winston, was "a pillar" of Olde Hampton's Bethel AME Church. Her sister Emily Winston received a citation from President Roosevelt, thanking her for more than one thousand hours of meritorious service as a nurse's aide during the war. The Winstons were the embodiment of the Double V, and Mary took her duties as secretary as seriously as if she were the head of the club.

Unsurprisingly, the USO was the scene of many a wartime romance. Negro soldiers from Fort Monroe and Langley Field and the naval training school on the campus of Hampton Institute rested their cares in the company of some of the community's most eligible bachelorettes. The USO's dance floor was always full of beautiful young ladies, but one enlistee stationed at the naval school had eyes only for the club's secretary. Mary's nimble intellect, her quiet but commanding nature, and her all-embracing humanitarian spirit might have been a red flag for a less secure man, but it was exactly her strength of character that drew Alabama native Levi Jackson to her. Their romance blossomed in the heady days of the war, and they married in 1944 at the Winston family home on Lincoln Street. Ever the independent spirit, Mary eschewed the traditional all-white bridal gown for a shorter white dress with black sequins, topped off with black gloves, black pumps, and a red rose corsage.

The end of the war brought the closing of the King Street USO and the end of Mary's job there. She worked for a brief period as a bookkeeper at Hampton Institute's Health Service but left after the birth of her son, Levi Jr., in 1946. While Levi Sr. headed off to work at his job as a painter at Langley Field, Mary doted on her son at home. With a full calendar of child care, family commitments, and volunteer activities, she was as busy as a stay-at-home mother as she had been working outside her home.

Her free time was absorbed by her position as the leader of Bethel AME's Girl Scout Troop No. 11. Scouting would be one of Mary's lifelong loves. The organization's commitment to preparing young women to take their place in the world, its mission to promulgate respect for God and country, honesty and loyalty—it was like a green-sashed version of all that Frank and Ella Winston had taught their children. Many of the girls in Troop 11 were from working-class, even poor, families—children of domestic servants, crab pickers, laborers—whose parents spent most of their waking hours trying to make ends meet. The door to the Jackson home on Lincoln Street was always open to them. Mary became a combination of teacher, big sister, and fairy godmother, helping her girls with algebra homework, sewing dresses for their proms, and steering them toward college.

Most of all, she went out of her way to provide them with the kinds of experiences that would expand their understanding of what was possible in their lives. With a leader as creative as Mrs. Jackson, Troop 11 was never inhibited by its modest resources. Rather than sitting down with the Girl Scout manual and going through the badge requirements as if it were simply a weekend version of a social studies class, she turned working toward those cheerfully embroidered green patches into an adventure, taking them on three-mile "country" hikes in local parks or field trips to the crab factory to learn more about what their parents did for work. For the hospitality badge, Mary arranged for the troop to attend an afternoon tea at the Hampton Institute Mansion House, a grand residence now occupied for the first time by a black president, Alonzo G. Moron. Mrs. Moron received the girls in high style, attended by a staff composed of students from the school's Home Economics Department. It was a sight the girls never forgot: an impeccable black staff in a fabulous house, serving a well-heeled *black* family. Not even the movies could compare with the glamour of that afternoon.

Once at a troop meeting at Bethel AME, Mary was leading her charges in a rendition of the folk tune "Pick a Bale of Cotton," complete with a pantomime of a slave working in the fields. It was a well-worn tune, one that she had sung before without much consideration. That day, however, the lyrics (*"We're gonna jump down, turn around, pick a bale of cotton!"*) and the shucky-jivey routine that accompanied it struck her like a bolt of lightning.

"Hold on a minute!" she said suddenly, interrupting the performance in mid-verse. The girls watched Mrs. Jackson, startled. Mary stood silent for a long moment, as if hearing the song for the first time. "We are never going to sing this again," she told them, trying to explain her reasoning to the surprised youngsters. The song reinforced all the crudest stereotypes of what a Negro could do or be. Sometimes, she knew, the most important battles for dignity, pride, and progress were fought with the simplest of actions.

It was a powerful moment for the girls of Troop 11. Mary didn't have the power to remove the limits that society imposed on her girls, but it was her duty, she felt, to help pry off the restrictions they might place on themselves. Their dark skin, their gender, their economic status— none of those were acceptable excuses for not giving the fullest rein to their imaginations and ambitions. You can do better—*we* can do better, she told them with every word and every deed. For Mary Jackson, life was a long process of raising one's expectations.

When Levi Jr. turned four, Mary Jackson filed an application with the Civil Service, applying both for a clerical position with the army and as a computer at Langley. In January 1951, she was quickly called up to work at Fort Monroe as a clerk typist. The job involved typing, filing, distributing mail, making copies—nothing more exotic than her previous work, but because of the sensitive nature of the documents that passed through the office, she was required to get a secret security clearance. The United States' anxiety over the threat posed by the Soviet Union had increased steadily since the end of World War II, and escalated in 1949 when the USSR detonated its first atomic bomb. One of the documents that circulated at Fort Monroe was an army plan to be executed in the case of an atomic attack.

The rivalry between the onetime allies exploded into open war-by-proxy along the border between North and South Korea in 1950, making the stakes of the new conflict concrete for most Americans, and for the NACA. Over Korean skies, Russian fighter planes "too fast to be identified"—the near supersonic MIG-15—attacked American B-29 Superfortress planes. "Russia Said to Have Fastest Fighter Plane," ran the headline of a 1950 *Norfolk Journal and Guide* article. The Americans had led the way through the sound barrier with Chuck Yeager at the helm of the X-1, but by 1950, the NACA reckoned that "the Russians expended at least three times the man power in their research establishments" than what the United States budgeted. Again the NACA angled to benefit from increased international tension, handing Congress a proposal to double its agency-wide employment level from seven thousand in 1951 to fourteen thousand in 1953.

The long list of job vacancies published in the *Air Scoop* was reminiscent of the boom times of the last war and buttressed America's vow that it would not back down before any rival in the heavens. With its many new facilities coming into operation, the laboratory again cast its net for the female alchemists who could turn the numbers from testing into aeronautical gold. With Mary's abilities, it was no surprise that Uncle Sam decided she would be of better use as an NACA computer than as a military secretary. After three months at Fort Monroe, she accepted an offer to work for Dorothy Vaughan.

In the eight years that had elapsed since Dorothy Vaughan had taken the same trip on her first day of work, the fields and remaining forest of Langley's West Side had filled in with roads, sidewalks, and the laboratory's characteristic low red brick buildings, an aeronautical village brimming with inhabitants. A gigantic 295-by-300-foot hangar, also known as Building 1244, the largest structure of its kind in the world, sheltered the laboratory's fleet of research aircraft, including the X plane series, the offspring of Chuck Yeager's sound-barrier-breaking X-1. The feat of smashing through the sound barrier scored a Collier Trophy, the aeronautics industry's most prestigious award, for Yeager,

Lawrence Bell—whose company, Bell Aircraft, produced the X-1—and John Stack, Langley's assistant director, who had championed the plane's development as a research tool. More importantly, breaching that physical barrier opened researchers' minds to the wider spectrum of powered flight's possibilities, and its challenges. As a plane accelerated from high subsonic speeds to low supersonic speeds, passing through the unsteady "transonic" region between Mach .8 and Mach 1.2, the simultaneous presence of subsonic and supersonic flows caused buffeting and instability. Aerodynamicists sharpened their pencils to understand the sudden changes in lift and drag on a plane flying at transonic speeds, because the transonic regime served as the waiting room for any vehicle seeking to supersede the speed of sound. The telltale sonic boom indicated that the plane had pushed through the volatile transonic region into the state of smoother, all-supersonic flows.

With Mach 1 achieved, engineering imaginations broke free of all previous speed limits. While maintaining its efforts to wring out improvements in subsonic flight and address the complexities of transonic flight, the NACA mounted a concerted effort to take what it had learned from the experimental planes and use it to design military production aircraft capable of supersonic flight. "For America to continue its present challenged supremacy in the air will require that it develop tactical military aircraft that will fly faster than sound before any other nation does so," John Victory, the NACA's long-serving executive secretary, said in the *Journal and Guide* article. The most visionary of the brain busters pined for the day when a pilot could take one of their creations for a hypersonic joy ride: Mach 5 or faster. The details of something mysteriously known only as Project 506 was revealed in 1950 to be a hypersonic wind tunnel, with a test section of just eleven inches, but capable of subjecting models to wind speeds close to Mach 7. That test facility, and a large complex under construction called the Gas Dynamics Laboratory, which would be capable of wind tunnel tests up to Mach 18, tipped the agency's interest in flight so fast that it could occur only at the limits of Earth's atmosphere. The vacuum spheres being built to power the tests in the Gas Dynamics Laboratory—three smooth-metal sixty-foot-diameter globes and a one-hundred-foot corrugated sphere towering over its siblings—would become one of the most recognizable landmarks on the Virginia Peninsula.

The same day that Mary Jackson started her job at Langley—April 5, 1951—a New York federal court handed down a death sentence against Ethel and Julius Rosenberg, a New York couple accused of spying for the Russians. The Cold War wasn't happening just in the skies above Korea or in a Europe that was being divided into a Soviet-allied East and a US-friendly West. The Rosenberg trial sparked fears inside the United States that living throughout the country were communist sympathizers plotting to overthrow the government. Official propaganda films like *He May Be a Communist* warned Americans that their neighbors might have thrown in their lot with the Reds. Even friends and family could be secret Communists, alerted the film, the kind "who don't show their real faces." The Rosenberg trial was all the evidence many citizens needed that their country had been infiltrated by radicalized agents of the Soviet Union.

At Langley, the Rosenberg trial and its repercussions hit a little too close to home. An engineer named William Perl, who had worked at Langley until he transferred to the NACA laboratory in Cleveland in 1943, was accused of stealing classified NACA documents and funneling them to the Soviet Union via the Rosenbergs. Among the secrets allegedly leaked by Perl were plans for a nuclear-powered airplane and the specs for a high-speed NACA airfoil. Some even believed that the high T-shaped tails on the MIGs that were shooting down American pilots over Korea were based on NACA designs. Perl was eventually tried and cleared of the espionage charges, but he was convicted of perjury for lying about his association with the Rosenbergs.

The FBI had begun laying the groundwork for the case in the late 1940s, interrogating Langley employees about their knowledge of Perl and his possible conspirators. Federal agents terrified staffers by showing up unannounced at their homes in Hampton and Newport News, ringing the doorbell in the evenings to ask questions. The FBI tracked down former Langley engineer Eastman Jacobs, known for his left-leaning sympathies, and interrogated him at his new home in California. They spent hours questioning Pearl Young, who had left the agency in the late 1940s for a job teaching physics at Penn State. The Stability Research Division, where Dorothy Hoover worked, was a particular target, as Perl had been a member of the group before leaving for Cleveland.

The investigation tapped into veins of anti-Semitism that flowed just under the racial prejudice at the laboratory and in the community. Quietly, some laboratory employees complained about the "New York communist people" and the "practically impossible New York Jews" recruited to work at Langley. A Jewish computer who had invited her Negro college roommate down to Virginia for a weekend visit caused a scandal. The progressives of the Stability Research group, regardless of their actual political practices, were certainly open to accusations of subversion for their embrace of "dangerous" ideas like racial integration, civil rights, and equality for women.

Investigators looked into rumors that engineers in Stability Research and a "black computer" with whom they were friendly had been caught burning the loyalty forms President Truman had required all civil servants to sign after 1947. In 1951, *Air Scoop* published a long list of organizations that the government had labeled totalitarian, Communist, or subversive, the clear message that affiliation with any of them might jeopardize one's job. Around the same time, Dorothy Vaughan's relative, Matilda West, possibly the black computer accused of disloyalty, was fired from her job at the laboratory. West was an outspoken advocate for black empowerment and one of the leaders of the local NAACP. The NAACP wasn't included on the government list, but it had long been a target of the Wisconsin senator Joseph McCarthy. With the Rosenberg trial casting a shadow on the NACA and its security practices, and with the agency's growing budget requests under the microscope in Congress, the lab's administrators may have decided that having a "radical" black computer on staff was a headache they just didn't need.

It was a dismissal that would shake West Computing to its core, with possibly career-damaging implications for Dorothy Vaughan as well. The Red scares and Communist hysteria of the late 1940s and early 1950s destroyed reputations, lives, and livelihoods, as Matilda West's situation proved. The fear of Communism was a bonanza for segregationists like Virginia senator Harry Byrd. Byrd painted the epithet "Communist" on everyone and everything that threatened to upend his view of "traditional" American customs and values, which included white supremacy. (One sequence in the film *He May Be a Communist* not so subtly showed a dramatized protest march in which participants held signs reading END KKK TERROR and NO WAR BASES IN AFRICA.)

Having the courage to criticize the government carried serious risks, and once again, the champions of Negro advancement had to engage in the delicate two-step of denouncing America's foreign enemies while doing battle with their adversaries at home. Even A. Philip Randolph, an avowed socialist who preached a fiery sermon in favor of fair employment and civil rights legislation in front of a packed audience in Norfolk in 1950, was careful in his speeches to denounce Communism as antithetical to the interests of the Negro people.

Paul Robeson, Josephine Baker, and W. E. B. Du Bois were among the black leaders to draw a connection between America's treatment of its Negro citizens and European colonialism. They traveled abroad and made speeches declaring their solidarity with the peoples of India, Ghana, and other countries that were in the early days of new regimes as independent nations or pushing with all their might to get there by agitating against their colonial rulers. The US government went so far as to restrict or revoke these firebrands' passports, hoping to blunt the impact of their criticism of American domestic policy in the newly independent countries that the United States was eager to persuade to join its side in the Cold War.

Foreigners who traveled to the United States often experienced the caste system firsthand. In 1947, a Mississippi hotel denied service to the Haitian secretary of agriculture, who had come to the state to attend an international conference. The same year, a restaurant in the South banned Indian independence leader Mahatma Gandhi's personal doctor from its premises because of his dark skin. Diplomats traveling from New York to Washington along Route 40 were often rejected if they stopped for a meal at restaurants in Maryland. The humiliations, so commonplace in the United States that they barely raised eyebrows, much less the interest of the press, were the talk of the town in the envoys' home countries. Headlines like "Untouchability Banished in India: Worshipped in America," which appeared in a Bombay newspaper in 1951, mortified the US diplomatic corps. Through its inability to solve its racial problems, the United States handed the Soviet Union one of the most effective propaganda weapons in their arsenal.

Newly independent countries around the world, eager for alliances that would support their emerging identities and set them on the path to long-term prosperity, were confronted with a version of the same question black Americans had asked during World War II. Why would a black or brown nation stake its future on America's model of democracy when within its own borders the United States enforced discrimination and savagery against people who looked just like them?

The international audience, and their opinion of US racial problems, were beginning to matter—a lot—to American leaders, and concern for their opinion influenced Truman's 1947 decision to desegregate the military through Executive Order 9981. At the start of the Korean War, the Tan Yanks remaining in active service in the US Air Force were called up to serve as a part of an integrated squadron.

At the same time, Truman issued Executive Order 9980, sharpening the teeth of the wartime mandate that had helped bring West Area Computing into existence. The new law went further than the measure brought to life by A. Philip Randolph and President Roosevelt by making the heads of each federal department "personally responsible" for maintaining a work environment free of discrimination on the basis of race, color, religion, or national origin. The NACA appointed a fair employment officer to enforce the measure and settled into the habit of responding to a quarterly questionnaire, relating its activity with respect to its growing numbers of black professional employees.

"The laboratory has one work unit composed entirely of Negro women, the West Area Computers, which may fall into the category of a segregated work unit," wrote Langley's administrative officer, Kemble Johnson, in a 1951 memo. "However, a large percentage of employees are usually detailed to work in non-segregated units for periods of one week to three months. Members of this unit are frequently transferred to other research activities at Langley, where they are integrated into non-segregated units. The same promotional activities are available to the West Area Computers as to other computers at Langley."

Supersonic aircraft and missiles were determining the course of the Cold War, but so too would "science textbooks and racial relations." The West Area Computers were ammunition for both fronts of the conflict, yet they were one of the best-kept secrets in the federal government. Among the middle-class and black professional community of southeastern Virginia, however, word traveled like wildfire: Mrs. Vaughan's office is hiring. Christine Richie heard about West Computing in the Huntington High School teachers' lounge. Aurelia Boaz, a 1949 Hampton Institute graduate, got the word through the college grapevine. It seemed that every black church on the peninsula had at least one member who worked out at Langley. Applications were passed along at homecoming tailgates and choir rehearsals, at the meetings of Delta Sigma Theta and Alpha Kappa Alpha sororities and the Newsome Park PTA. Mary Jackson was connected to so many computers in so many different ways that the only surprising thing about her arrival at Langley was that it had taken so long.

CHAPTER ELEVEN

The Area Rule

In the early 1950s, there was rarely a slow workday for Dorothy Vaughan. With research activity concentrated on the West Side of Langley's campus, Dorothy managed a steady stream of computing jobs, dispatching incoming assignments to the women in her office and sending her computers out to various engineering groups located in the vicinity with greater frequency. Most of the work that came into the West Area computing office originated from one of the tunnels on the West Side or from the Flight Research Division, which was located in Building 1244, the West Side's new hangar. Though the East Side was now smaller in size and activity than Langley's booming West Side, facilities like the Spin Tunnel (a building shaped like a squat smokestack where engineers analyzed models subjected to dangerous spins) and Tanks nos. 1 and 2 (three-thousand-foot-long channels for testing seaplanes) remained busy. The Full-Scale Tunnel, the linchpin of the lab's World War II drag cleanup work, continued to test everything from low-speed aircraft designed with delta wings to helicopters. During intense periods, if the work exceeded the available hands, computing supervisors working in the East Area might put in a phone call to Dot Vaughan for reinforcements.

On one such occasion, two years after Mary joined West Computing, Dorothy Vaughan sent Mary to the East Side, staffing her on a project alongside several white computers. The routines of the computing work had become familiar to Mary, but the geography of the East Side was not. Her morning at the East Side job proceeded without incident—until nature called.

"Can you direct me to the bathroom?" Mary asked the white women.

They responded to Mary with giggles. How would *they* know where to find *her* bathroom? The nearest bathroom was unmarked, which meant it was available to any of the white women and off-limits to the black women. There were certainly colored bathrooms on the East Side, but with most black professionals concentrated on the West Side, and fewer new buildings on the East Side, Mary might need a map to find them. Angry and humiliated, she stormed off on her own to find her way to *her* restroom.

Negotiating racial boundaries was a daily fact of Negro life. Mary wasn't naive about the segregation at Langley—it was no different than anywhere in town. Yet she couldn't shake this particular incident. It was the proximity to professional equality that gave the slight such a surprising and enduring sting. Unlike the public schools, where minuscule budgets and ramshackle facilities exposed the sham of "separate but equal," the Langley employee badge supposedly gave Mary access to the same workplace as her white counterparts. Compared to the white girls, she came to the lab with as much education, if not more. She dressed each day as if she were on her way to a meeting with the president. She trained the girls in her Girl Scout troop to believe that they could be anything, and she went to lengths to prevent negative stereotypes of their race from shaping their internal views of themselves and other Negroes. It was difficult enough to rise above the silent reminders of Colored signs on the bathroom doors and cafeteria tables. But to be confronted with the prejudice so blatantly, there in that temple to intellectual excellence and rational thought, by something so mundane, so ridiculous, so universal as having to go to the bathroom ... In the moment when the white women laughed at her, Mary had been demoted from professional mathematician to a second-class human being, reminded that she was a black girl whose piss wasn't good enough for the white pot.

Still fuming as she walked back to West Computing later that day, Mary Jackson ran into Kazimierz Czarnecki, an assistant section head in the Four-by-Four-Foot Supersonic Pressure Tunnel. A stocky fellow with a lantern jaw who played first base in the Langley softball league, Kazimierz Czarnecki—friends called him "Kaz"—was a native of New Bedford, Massachusetts, who had come to the laboratory in 1939 after graduating from the University of Alabama with an aeronautical engineering degree. His good nature and prodigious research output made him a well-liked, well-respected member of the laboratory staff. Before joining the Four-Foot SPT, Kaz had worked as a member of the Nine-Inch Supersonic Tunnel research staff, which maintained an office in the Aircraft Loads Building, where West Area Computing was housed.

Most blacks automatically put on a mask around whites, a veil that hid the "dead-weight of social degradation" that scholar W. E. B. Du Bois gave voice to so eloquently in *The Souls of Black Folk*. The mask offered protection against the constant reminders of being at once American, and the American dilemma. It obscured the anger that blacks knew could have life-changing—even life-ending—consequences if displayed openly. That day, however, as Mary Jackson ran into Kazimierz Czarnecki on the west side of the Langley Aeronautical Laboratory, there was no turning inward, no subversion, and no dissembling. Mary Jackson let her mask slip to the ground and answered Czarnecki's greeting with a Mach 2 blowdown of frustration and resentment, letting off steam as she ranted about the insult she had experienced on the East Side.

Mary Jackson was a soft-spoken individual, but she was also forth-right and unambiguous. She chose to speak to everyone around her in the same serious and direct fashion, whether they were adolescents in her Girl Scout troop or engineers in the office. Mary was also a shrewd and intuitive judge of character, an emotionally intelligent woman who paid close attention to her surroundings and the people around her.

Whether her outpouring in front of Czarnecki was the spontaneous result of having reached a breaking point or something more astute, she picked the right person to vent to. What had started as one of the worst workdays Mary Jackson had ever had would end up being the turning point of her career.

"Why don't you come work for me?" Czarnecki asked Mary. She didn't hesitate to accept the offer.

While the national press published stories on Langley's link to the Rosenberg scandal, industry outlets like *Aviation Week* lauded the laboratory for two related advances that would revolutionize high-speed aircraft production: slotted walls in wind tunnels and an innovation known as the Area Rule.

The point of a wind tunnel, of course, was to simulate as closely as possible the conditions that prevailed in free flight. Interference from airflows bouncing off the solid walls of the test section, one of the phenomena examined by Margery Hannah and Sam Katzoff in their 1948 report, was one of the limitations of ground-based testing. The problem was most notable in the transonic range, as the eddies of air surrounding an object approached the speed of sound. A Langley researcher named Ray Wright had the intuition that cutting holes or slots in the walls of the wind tunnels would alleviate the interference effects, a concept that was proven when Langley built a small test tunnel with perforated walls. In 1950, they retrofitted the Sixteen-Foot High-Speed Tunnel (rechristened the Sixteen-Foot Transonic Dynamics Tunnel) with slotted walls and then did the same for the Eight-Foot High-Speed Tunnel. Taming the tunnel interference was a "long sought technical prize" for the researchers, and in 1951 it earned John Stack and his colleagues another coveted Collier Trophy.

The new tunnel design set the stage for the second of the decade's significant developments. An engineer named Richard Whitcomb noticed that in the transonic speed range, the greatest turbulence occurred at the point where the wings of a model plane connected to its fuselage. Indenting the plane's body inward along that joint reduced the drag dramatically and resulted in an increase of as much as 25 percent in the plane's speed for the same level of power. The Area Rule (so-called because the formula predicted the correct ratio of the area of a cross-section of a plane's wing to the area of the cross-section of its body) had the potential to have a greater impact on everyday aviation than supersonic aircraft, because of the thousands of aircraft whose operating speed topped out at the transonic range. The press had more than the usual fun with such an esoteric engineering concept, calling the new planes "wasp-waisted" and "Coke-bottle shaped" and talking about the "Marilyn Monroe effect." Whitcomb scored a sit-down with CBS news anchor Walter Cronkite and gained a measure of local celebrity ("Hampton Engineer Besieged by Public," read a somewhat hyperbolic *Daily Press* headline). In 1954, Whitcomb would take home Langley's third Collier Trophy in less than a decade.

For all the advances that had occurred on the laboratory's watch since 1917—cowled engines, laminar flow airfoils, supersonic research planes, an icing tunnel that led to improvements in flight safety in freezing temperatures—the existing body of aeronautical knowledge still sheltered unexplored corners. The investments in new and upgraded facilities on Langley's West Side made in the late 1940s and early 1950s were yielding research breakthroughs and impacting the nature of the assignments Dorothy handed out to her staff.

Unlike academically oriented research organizations, the NACA's laboratories always strove to live up to the "practical solutions" of their founding mission. The hands-on nature of the work at Langley was visible in the planes parked in the hangar, in the workshops where craftsmen built models to the engineers' specifications, in the work of the mechanics who affixed the models in the proper positions in the test sections, and in the guts of the powerful new tunnels like the Unitary Plan Tunnel, which looked like "an oil refinery under a roof." No matter how abstract the work or how conceptual the problem being solved, no one at Langley ever forgot that behind the numbers was a real-world goal: faster planes, more efficient planes, safer planes.

·   ·   ·

Of course, the NACA wasn't such a bad place for the theoretical engineers either. Dorothy Hoover thrived in the Stability Analysis Division. By 1951, she had earned the lofty title of aeronautical research scientist, graded GS-9 in the government's revamped rating system. When Hoover's boss, R. T. Jones, left Langley for the NACA's Ames laboratory in 1946, Dorothy continued her work with the group's other notable researchers. Her Langley career reached a peak in 1951 with the publication of two reports, one with Frank Malvestuto, the other with Herbert Ribner, both of them detailed analyses of the swept-back wings that were now a standard feature of production aircraft. What the compressed-air and fresh-air engineers examined through direct observation, the theoreticians approached through fifty-page treatises in which one single equation might occupy the better part of a page. If research production was a measure of career viability—and it was— theoretical aerodynamics might have been the best place in the world to be a female researcher. Dorothy Hoover, Doris Cohen, and at least three other women published one or more reports with the group between 1947 and 1951. The leaders of the group clearly valued and cultivated the talent of their female members. Perhaps it was the remove from the brawnier aspects of engineering that made the theoretical group such a productive environment for women.

In 1952, Dorothy Hoover decided to take a leave of absence from the world of engineering and give herself over to the theoretical pursuits that were closest to her heart. She resigned from Langley and returned to her alma mater, Arkansas AM&N, for a master's degree in mathematics. Her thesis, "On Estimates of Error in Numerical Integration," was included in the 1954 proceedings of the Arkansas Academy of Science. That same year she enrolled in the University of Michigan under a John Hay Whitney Fellowship, a program designed to match talented Negro scholars with the country's most competitive graduate programs.

Mary Jackson, on the other hand, leaned in to the the engineer's paradise that was the NACA. With a background in math and physics, she brought to the job an understanding of the physical phenomena behind the calculations she worked on. And the Langley people were busy people like her, running off after work to play on one of the laboratory's sports teams or attend a club meeting or lecture. Many of them tutored kids in math and science, something that Mary had done since graduating from college. Whether or not she had it planned at the time, Mary Jackson was on her way to becoming a Langley lifer.

During new-employee induction on her first day of work, Mary Jackson had met James Williams, a twenty-seven-year-old University of Michigan engineering grad and former Tuskegee airman who had fallen in love with airplanes as a teenager. Williams applied for engineering positions through the Civil Service but had been wary of moving to a state south of the Mason-Dixon Line. Langley's personnel officer, Melvin Butler, courted Williams energetically by phone, trying to convince him to to accept the laboratory's offer. He even made arrangements for a place for Williams to live in Hampton. Further enticement was provided by a beautiful psychology grad student named Julia Mae Green, who after graduation would be returning to her native Virginia. Butler, perhaps trying to circumvent complaints that might short-circuit his offer, did not disclose ahead of time to Langley's engineering staff that the newest recruit was black. Williams wasn't the first black engineer hired by Langley, but the couple of black men who preceded him had come and gone so quickly that not even their names remained in the institutional memory.

On his first day, Williams had had to convince the guards at the Langley security gate that he wasn't a groundskeeper or cafeteria worker so that he could get passed along for processing as an engineer. Several white supervisors refused him a place in their groups, but an influential branch chief in the Stability Research Division named John D. Bird—"Jaybird"—raised his hand right away to offer the young man a position. "Jaybird was as fair as it got," Williams's wife, Julia, remembered years later.

Not everyone in the group was as enthusiastic as Bird. "So how long do you think you're going to be able to hang on?" one new office mate teased, referring to the black engineers who had washed out. "Longer than you!" Williams retorted. Whereas the black women enjoyed the support that came from being part of a group, starting in a pool was not an option for a male engineer. Williams and the other black men who were soon to follow in his footsteps had a more solitary work life and faced aggressions that the women did not. But even though it was the black women who broke Langley's color barrier, paving the way for the black men now being hired, the women would still have to fight for something that the black men could take for granted: the title of engineer.

Soon after moving to the Four-by-Four-Foot Supersonic Pressure Tunnel, Mary Jackson was given an assignment by John Becker, the chief of the Compressibility Division (compressibility referred to the compression of air molecules characteristic of faster-than-sound flight), Kazimierz Czarnecki's boss's boss's boss. Langley liked to think of itself as a place that eschewed bureaucracy, where an idea from a cafeteria worker could get a fair hearing if it were compelling enough. Division chiefs, just two rungs down from the top post in the laboratory, however, were Very Important People. John Becker, heir to John Stack and Eastman Jacobs and other NACA legends, ruled an empire composed of the Four-Foot SPT and all the other tunnels devoted to supersonic and hypersonic research. Becker was the kind of guy the eager front-row boys from the top engineering programs would do anything to impress.

John Becker gave Mary Jackson the instructions for working through the calculations. She delivered the finished assignment to him just as she completed her work for Dorothy Vaughan, double-checking all numbers, confident that they were correct. Becker reviewed the output, but something about the numbers didn't seem right to him. So he challenged Mary's numbers, insisting that her calculations were wrong. Mary Jackson stood by her numbers. She and her division chief went back and forth over the data, trying to isolate the discrepancy. Finally, it became clear: the problem wasn't with her output but with his input. Her calculations were correct, based on the wrong numbers Becker had given her.

John Becker apologized to Mary Jackson. The episode earned Mary a reputation as a smart mathematician who might be able to contribute more than just calculations to her new group. Her showdown with John Becker was the kind of gambit that the laboratory expected, encouraged, and valued in its promising male engineers. Mary Jackson—a former West Computer!—had faced down the brilliant John Becker and won. It was a cause for quiet celebration and behind-the-scenes thumbs-up among all the female computers.

Most engineers were also good mathematicians. But it was the women who massaged the numbers, swam in the numbers, scrutinized the numbers until their eyes blurred, from the time they set their purses down on the desks in the morning until the time they put on their coats to leave at the end of the day. They checked each other's work and put red dots on the data sheets when they found errors—and there were very, very few red dots. Some of the women were capable of lightning-fast mental math, rivaling their mechanical calculating machines for speed and accuracy. Others, like Dorothy Hoover and Doris Cohen, had highly refined understandings of theoretical math, differentiating their way through nested equations ten pages deep with nary an error in sign. The best of the women made names for themselves for accuracy, speed, and insight. But having the independence of mind and the strength of personality to defend your work in front of the most incisive aeronautical minds in the world—that's what got you noticed. Being willing to stand up to the pressure of an opinionated, impatient engineer who put his feet up on the desk and waited while you did the work, who wanted his numbers done right and done yesterday, to spot the bug in his logic and tell him in no uncertain terms that he was the one who was wrong—that was a rarer quality. That's what marked you as someone who should move ahead.

CHAPTER TWELVE

Serendipity

It had always been Katherine Goble's great talent to be in the right place at the right time. In August 1952, twelve years after leaving graduate school, that right place was back in Marion, the site of her first teaching job, at the wedding of her husband, Jimmy Goble's, little sister Patricia. Pat, a vivacious college beauty queen just two months graduated from Virginia State College, was marrying her college sweetheart, a young army corporal named Walter Kane.

Katherine and Jimmy packed the girls into the car and drove the sixty miles from Bluefield to Jimmy's parents' house, which quivered with excitement for Pat's big day. Hotels in the South denied service to black patrons; blacks of all social strata knew to make arrangements with friends and family, or even strangers known for opening their homes to guests, rather than risk embarrassment or possibly danger while traveling. Five of Jimmy's eleven siblings still lived in Marion, and their houses swelled to maximum capacity accommodating the out-of-town visitors, including the groom's family, in from nearby Big Stone Gap, Virginia, and friends and extended family from all over Virginia, West Virginia, and North Carolina.

The simple but elegant wedding took place at the home of Jimmy's eldest sister, Helen. Pat, radiant in a ballerina-length accordion-pleated gown, stood before the makeshift altar adorned with evergreens and gladioli and said "I do" to Walter, dashing in a white dinner jacket. The jubilant crowd toasted the newly minted Mr. and Mrs. Kane. Katherine and Jimmy danced and ate wedding cake. Their three girls—Joylette, age eleven; Connie, ten; and Kathy, nine—squealed with delight as they played hide-and-seek and hopscotch and danced with their cousins. The celebration continued late into the cool August night and trickled into the following day, as the Goble and Kane clans tarried to savor their last moments together before heading back to their workaday lives.

Jimmy's sister and brother-in-law, Margaret and Eric Epps, had journeyed from Newport News, and the newlyweds planned to accompany the Eppses back to the coast, hitching a ride to their honeymoon at Hampton's segregated Bay Shore Beach resort. "Why don't y'all come home with us too?" Eric asked Katherine. "I can get Snook a job at the shipyard," he said, using Jimmy's family nickname. "In fact, I can get both of you jobs." There's a government facility in Hampton that's hiring black women, Eric told Katherine, and they're looking for mathematicians. It's a civilian job, he told her, but attached to Langley Field, in Hampton.

Jimmy's brother-in-law was the director of the Newsome Park Community Center. Since 1943, Eric Epps had coordinated community activities such as the semipro Newsome Park Dodgers baseball team, and he had been a strong advocate for the residents of his neighborhood with the local government, always fending off the demolition campaigns that never seemed to disappear. A former teacher in Newport News's public schools, he had taken the Newsome Park job after being fired for joining what was one of the most bitter and contentious of all of Virginia's teacher-pay-equalization lawsuits. Through his job and relationships with the residents in Newsome Park, he was one of the most well-connected individuals on the Virginia Peninsula, and knew many women of West Computing, including Dorothy Vaughan, who lived in the neighborhood.

Katherine listened intently as her brother-in-law described the work, her thumb cradling her chin, her index finger extended along her cheek, the signal that she was listening carefully. She and Jimmy made a living as public school teachers, but their paychecks were modest. The needs of their three growing daughters seemed greater by the day, and Katherine pushed her math skills to the breaking point just to cover their basics and squeeze out a little extra for piano lessons or Girl Scouts. Deft with a sewing machine, Katherine bought fabric from the dry goods store and stayed up nights making school outfits for the girls and dresses for herself. During the summer break from school, the Gobles worked as live-in help for a New York family that summered in Virginia's Blue Ridge Mountains. The extra cash tided them through the toughest months during the rest of the year.

Katherine enjoyed teaching. She felt a keen sense of responsibility to "advance the race," instilling not just book knowledge but discipline and self-respect in her students, who would need both of those qualities in abundance to make their way in a society stacked against them in nearly every way. She and Jimmy hewed to the path that was so well-worn that the feet of Negro college graduates like them were trained to walk it almost automatically. But Eric Epps's mention of the mathematics job in Hampton kindled the memory of a long dormant ambition, one Katherine was surprised to find still smoldering within her.

It was late that night when Katherine and Jimmy tucked the girls into their beds and collapsed onto their own bed, but they stayed up even later laughing and gossiping, recapping the family gathering. Only after exhausting all the other stories of the wedding did they broach the topic that occupied both of their minds. Taking the road to Newport News would mean making a decision quickly. With the school year approaching, the Bluefield principal would need time to find replacements to take over their classes. They would need a place to live. Uprooting the girls on short order and enrolling them in new schools would be trying for all of them. Hampton Roads was far from Jimmy's parents, and even farther away from Katherine's parents in White Sulphur Springs, who doted on their granddaughters. In the mountains, even the summer nights were breezy. How would she manage that coastal heat? It would have been easy to continue with the stable small-town life they had created. But the possibility of this new job tugged at the curiosity that was at the core of Katherine Goble's nature.

"Let's do it," she whispered.

Within one busy year, Katherine Goble and her family managed to fold themselves seamlessly into the Peninsula's rhythms. Newsome Park was a natural place for the quintet to land, its endless city-within-a-city blocks a ready-made brew of neighbors, social organizations, and advice for newcomers. Eric Epps had made good on his promise to find employment for his brother-in-law Jimmy Goble, who had traded his teaching credential for a painter's job at the Newport News shipyard. It was the kind of stable, well-paid job that gave Negro men—even those with white-collar credentials—a chance to pull their families solidly into the middle class. The girls loved their new school and marveled at living in such a large and dynamic black community.

Langley's personnel department approved Katherine's 1952 application, but with a June 1953 appointment. The intervening year helped her make the transition to everything but the Virginia heat; many was the night she pined for White Sulphur Springs' brisk summer nights. Filling the months in the meantime posed not the slightest problem. As a substitute math teacher at Huntington High School in Newport News, she had the perfect platform for meeting families in the area, and the Twenty-Fifth Street USO Club, which had continued operations as a postwar community center, recruited their new neighbor to be the club's assistant director. From her involvement with the local chapter of her sorority, Alpha Kappa Alpha, and her choice of a church, Carver Presbyterian, Katherine gained a strong social network and a new best friend.

Eunice Smith lived three blocks down, and Katherine delighted in learning that her soror, neighbor, and fellow worshipper was also a nine-year West Computing veteran. In the early days of June 1953, when Eunice Smith drove over to Katherine's house to pick her up for work, the two women started a routine that would persist for the next three decades. They made small talk as they drove through the tidal flatland of Hampton, Katherine's cat-shaped, wire-rimmed glasses lending her face a seriousness that matched her demeanor.

The morning commute ended at Mrs. Vaughan's office in the Aircraft Loads Building. It was a great and pleasant surprise to find that Katherine's new boss was not just a fellow West Virginian but the neighbor who had spent so much time with her family back in White Sulphur Springs. It didn't take long for Katherine to appreciate Dorothy's talents both as a mathematician and as a manager. When they needed more computing power, engineers trusted Dorothy to staff the right person for the job, often hoping that she was at the top of her own list.

Matching ability with assignments was only part of the challenge. The more subtle management skill was to match temperaments with the groups. The engineers could be a quirky lot, often brusque, temperamental, or authoritarian. One girl's brusque was another's cruel. Working closely in a team was key to the entire operation, and Dorothy had both a license and an obligation to see to it that her computers were set off on the best career paths possible.

For two weeks, Katherine worked the desk, learning the ropes. Her honors degree in mathematics, her time in graduate school, and her years teaching math added up to the very modest job rating of SP-3: a level 3 subprofessional, the entry-level fate of most of the women hired at Langley, regardless of their professional and educational credentials. Nearly twenty years after Virginia Tucker first came to Langley, and despite the fact that hundreds of women had gone through the position, it was still expected that the women would accept their new jobs with a little gee-I'm-just-glad-to-be-here gratitude. "Don't come in here in two weeks asking for a transfer" to an engineering group, the human resources director said to Katherine Goble on her first day of work, a comment she didn't appreciate. But she nonetheless felt "very, very fortunate" to have lucked into a job that paid her three times her salary as a teacher.

In the first few days, Katherine caught on to the routine of filling in data sheets according to equations that had been laid out by Dorothy Vaughan or one of the engineers, who made regular appearances in the office throughout the day. Two weeks after Katherine arrived, when a white man in shirtsleeves came into the office and approached Doro-thy Vaughan's station, initiating a quiet conversation, Vaughan nodded her head and scanned the array of desks, scrutinizing their occupants as she listened. After he left, Vaughan called Katherine Goble and an-other woman, Erma Tynes, to her desk. "The Flight Research Division is requesting two new computers," Vaughan said. "I'm sending you two. You're going to 1244."

For Katherine, being selected to rotate through Building 1244, the kingdom of the fresh-air engineers, felt like an unexpected bit of for-tune, however temporary the assignment might prove to be. She had been elated simply to sit in the pool and calculate her way through the data sheets assigned by Mrs. Vaughan. But being sent to sit with the brain trust located on the second floor of the building meant getting a close look at one of the most important and powerful groups at the lab-oratory. Just prior to Katherine's arrival, the men who would be her new deskmates, John Mayer, Carl Huss, and Harold Hamer, had presented their research on the control of fighter airplanes in front of an audience of top researchers, who had convened at Langley for a two-day confer-ence on the latest thinking in the specialty of aircraft loads.

With just her lunch bag and her pocketbook to take along, Kath-erine "picked up and went right over" to the gigantic hangar, a short walk from the West Computing office. She slipped in its side door, climbed the stairs, and walked down a dim cinderblock hallway until she reached the door labeled Flight Research Laboratory. Inside, the air reeked of coffee and cigarettes. Like West Computing, the office was set up classroom-style. There were desks for twenty. Most of the people in the space were men, but interspersed among them a few women con-sulted their calculating machines or peered intently at slides in film viewers. Along one wall was the office of the division chief, Henry Pearson, with a station for his secretary just in front. The room hummed with pre-lunch activity as Katherine surveyed it for a place to wait for her new bosses. She made a beeline for an empty cube, sitting down next to an engineer, resting her belongings on the desk and offering the man her winning smile. As she sat, and before she could issue a greeting in her gentle southern cadence, the man gave her a silent sideways glance, got up, and walked away.

Katherine watched the engineer disappear. Had she broken some unspoken rule? Could her mere presence have driven him away? It was a private and unobtrusive moment, one that failed to dent the rhythm of the office. But Katherine's interpretation of that moment would both depend on the events in her past and herald her future.

Bemused, Katherine considered the engineer's sudden departure. The moment that passed between them could have been because she was black and he was white. But then again, it could have been because she was a woman and he was a man. Or maybe the moment was an interaction between a professional and a subprofessional, an engineer and a girl.

Outside the gates, the caste rules were clear. Blacks and whites lived separately, ate separately, studied separately, socialized separately, worshipped separately, and, for the most part, worked separately. At Langley, the boundaries were fuzzier. Blacks were ghettoed into separate bathrooms, but they had also been given an unprecedented entrée into the professional world. Some of Goble's colleagues were Yankees or foreigners who'd never so much as met a black person before arriving at Langley. Others were folks from the Deep South with calcified attitudes about racial mixing. It was all a part of the racial relations laboratory that was Langley, and it meant that both blacks and whites were treading new ground together. The vicious and easily identifiable demons that had haunted black Americans for three centuries were shape-shifting as segregation began to yield under pressure from social and legal forces. Sometimes the demons still presented themselves in the form of racism and blatant discrimination. Sometimes they took on the softer cast of ignorance or thoughtless prejudice. But these days, there was also a new culprit: the insecurity that plagued black people as they code-shifted through the unfamiliar language and customs of an integrated life.

Katherine understood that the attitudes of the hard-line racists were beyond her control. Against ignorance, she and others like her mounted a day-in, day-out charm offensive: impeccably dressed, well-spoken, patriotic, and upright, they were racial synecdoches, keenly aware that the interactions that individual blacks had with whites could have implications for the entire black community. But the insecurities, those most insidious and stubborn of all the demons, were hers alone. They operated in the shadows of fear and suspicion, and they served at her command. They would entice her to see the engineer as an arrogant chauvinist and racist if she let them. They could taunt her into a self-doubting downward spiral, causing her to withdraw from the opportunity that Dr. Claytor had so meticulously prepared her for.

But Katherine Goble had been raised not just to command equal treatment for herself but also to extend it to others. She had a choice: either she could decide it was her presence that provoked the engineer to leave, or she could assume that the fellow had simply finished his work and moved on. Katherine was her father's daughter, after all. She exiled the demons to a place where they could do no harm, then she opened her brown bag and enjoyed lunch at her new desk, her mind focusing on the good fortune that had befallen her.

Within two weeks, the original intent of the engineer who walked away from her, whatever it might have been, was moot. The man discovered that his new office mate was a fellow transplant from West Virginia, and the two became fast friends. West Virginia never left Katherine's heart, but Virginia was her destiny.

CHAPTER THIRTEEN

Turbulence

At six months and counting, Katherine Goble's temporary assignment in the Flight Research Division was starting to look terribly permanent. So at the beginning of 1954, Dorothy Vaughan made her way to Building 1244 for a sit-down with Henry Pearson, the head of the branch that had "borrowed" her computer and forgotten to return her.

Katherine's offer to begin work at Langley in 1953 had come with a six-month probational appointment. Successful completion of the trial period would make her eligible for promotion from the entry level of SP-3 to SP-5, with the raise that accompanied it. Though Katherine had spent only two weeks physically in the West Area Computing office, she was still Dorothy's responsibility. Katherine could have been classified as a permanent member of West Computing, like the rest of the women who reported to Dorothy, available to rotate through other groups on temporary assignments. Or Henry Pearson could make Katherine an offer to officially join his group, as Kazimierz Czarnecki had done with Mary Jackson. One way or the other, however, Dorothy Vaughan and Henry Pearson needed to resolve Katherine Goble's situation.

"Either give her a raise or send her back to me," Dorothy said to Henry Pearson, sitting upstairs in his office in 1244. A Langley engineer in the old style, Pearson had graduated from Worcester Polytechnic in

Massachusetts and started work at the lab in 1930. He was a keen golfer, a horn-rimmed glasses wearer, the epitome of the New England WASP. Pearson was not a big fan of women in the workplace. His wife did not work; rumor had it that Mrs. Henry Pearson had been forbidden by her husband from holding a job.

As a branch chief attached to the high-profile Flight Research Division, Henry Pearson stood a level above Dorothy in the Langley management hierarchy. By the time Dorothy came to the laboratory in 1943, Pearson had already served as an assistant division chief for many years. Fearless as she was, Dorothy would have approached Henry Pearson even if she weren't a manager, but the official title of section head lent her additional authority. It put her on equal footing with the other female supervisors and—theoretically, at least—with men of the same rating, and it afforded her a degree of center-wide visibility. When calculating machine manufacturer Monroe asked Langley for its help producing a handbook on how to work algebraic equations with its machines, Dorothy was drafted as a consultant, working on a team with other well-respected women at Langley, including Vera Huckel from the Vibration and Flutter branch and Helen Willey of the Gas Dynamics complex.

The meeting between Dorothy Vaughan and Henry Pearson ended as they both knew it would, with Pearson offering Katherine Goble a permanent position in his group, the Maneuver Loads Branch, with a corresponding increase in salary. Dorothy's insistence also had a collateral effect: one of the white computers in the branch, in the same limbo position as Katherine, had herself gone to Pearson to petition for a raise. The white woman's request had fallen on deaf ears. *The rules are the rules*, Dorothy reminded Henry Pearson. Dorothy wielded her influence to win promotions for both Katherine and her white colleague.

The fact was, the engineers who worked for Henry Pearson realized soon after Katherine Goble took a seat at her desk in 1244 that their new computer was a keeper, and they had no intention of sending her back. Katherine's familiarity with higher-level math made her a versatile addition to the branch. Her library of graduate-school textbooks crowded onto her desk next to the the the calculating machine, ready references if she needed them.

The Flight Research Division was a den of high-energy, free-thinking, aggressive, and very smart engineers. They and their brethren in the Pilotless Aircraft Research Division (PARD), a group specializing in the aerodynamics of rockets and missiles, spent their time not in the confines of the wind tunnels but in the company of live, fire-breathing, ear-splitting, temperamental metal projectiles. The "black-haired, leather-faced, crew-haircutted human cyclone" head of the Flight Research Division, Langley's chief test pilot, Melvin Gough, early in his engineering career had decided to take his life in his hands to train as a test pilot in order to improve the quality of his research reports. Testosterone filtered up from the hangar along with the jet-fuel fumes. It wasn't the kind of place that would exhibit particular patience for anyone, male or female, who took too long to scale the learning curve. Timidity in the Flight Research Division would get a girl nowhere.

Fortunately, Katherine Goble's confidence in her own mathematical abilities, and her innate curiosity, pushed her to pepper the engineers with questions, just as she had as a child with her parents and teachers. They fielded her inquiries with gusto: they could, and did, spend most of their lives talking and thinking about flight and would never run out of patience for the topic.

The Maneuver Loads Branch conducted research on the forces on an airplane as it moved out of stable, steady flight or tried to return to stable, steady flight. A sister branch, Stability and Control, developed the systems that would provide a plane with a smooth ride through rough air. The vehicles at the extreme experimental end of the aeronautical spectrum were the ones that made the romantic aeronautical engineer's heart beat faster—supersonic planes, hypersonic planes, planes capable of brushing the limits of space—but the transportation revolution fostered in no small part by Langley engineers like Henry Pearson had created a demand for research on vehicles designed for much more pedestrian pursuits. One of the tasks of the Maneuver Loads Branch was to examine safety concerns provoked by increasingly crowded skies.

One of the first assignments to land on Katherine's desk involved getting to the bottom of an accident involving a small Piper propeller plane. The plane, which was flying along in otherwise unremarkable fashion, literally fell out of the clear blue sky and crashed to the ground. The NACA received the plane's flight recorder, and the engineers assigned Katherine to analyze the photographic film record of the flight's vital signs, the first step in the search for answers as to what might have befallen the plane. For hours upon hours, day after day, she sat in a dark room and peered through a film reader, noting and writing down the airspeed, acceleration, altitude, and other metrics of the flight that were measured in regular time intervals over the course of the flight. The engineers specified any conversions to be applied to the raw data—converting, for example, miles per hour to feet per second—and supplied Katherine with the equations to be used to analyze the converted data. As a final step, Katherine plotted the data in order to give engineers a visual snapshot of the plane's disrupted flight.

Then the engineers set up an experiment re-creating the circumstances of the accident, flying a test plane into the trailing wake of a larger plane. The data from that, too, washed onto Katherine Goble's desk: seemingly endless hours, days, weeks, months of the same thing. It was typical eye-straining, monotonous computing work—and Katherine loved every moment of it.

When the engineers analyzed Katherine's reduced data, they were fascinated, realizing they were uncovering something they had not quite seen before. It turned out that the Piper had flown perpendicularly across the flight path of a jet plane that had just passed through the area. A disturbance caused by a plane could trouble the air for as long as half an hour after it flew through. The wake vortex of the larger plane had acted like an invisible trip wire: upon crossing the rough river of air left behind by the jet, the propeller plane stumbled in midair and tumbled out of the sky. That research, and other investigations like it, led to changes in air traffic regulations, mandating minimum distances between flight paths so as to prevent that kind of wake turbulence accident.

When Katherine Goble read the report, she found it "one of the most interesting things [she] had ever read," and felt tremendous satisfaction to have participated in something that would have positive, real-world results. Her enthusiasm for the work, even the parts that others considered drudgery, was irrepressible. She couldn't believe her good fortune, getting paid to do math, the thing that came most naturally to her in the world.

She took a genuine liking to her new colleagues as well. The West Virginia engineer she had met on day one played oboe in a local symphony. Members of the Brain Busters Club convened after work and on weekends to build elaborate model airplanes by hand. Many of the men and women at Langley joined softball or basketball teams and played in local amateur leagues. Langley's "Skychicks" competed against a team fielded by the power company, the Kilowatt Cuties; over time, the black employees joined teams as well. And then there was the lunchtime bridge game. The game's requirement of both analytical and people-reading skills made it a favorite of the engineers, and they spent many a lunch hour in fierce competition. They were an opinionated, high-energy bunch, and best of all, as far as Katherine was concerned, they were all as smart as whips. There was nothing Katherine Goble loved more than *brains*.

From the very beginning, Katherine felt completely at home at Langley. Nothing about the culture of the laboratory or her new office rattled her—not even the persistent racial segregation. At the beginning, in fact, she didn't even realize the bathrooms were segregated. Not every building had a Colored bathroom, a fact that Mary Jackson had discovered so painfully during her rotation on the East Side. Though bathrooms for the black employees were clearly marked, most of the bathrooms—the ones implicitly designated for white employees—were unmarked. As far as Katherine was concerned, there was no reason why she shouldn't use those as well. It would be a couple of years before she was confronted with the whole rigmarole of separate bathrooms. By then, she simply refused to change her habits—refused to so much as enter the Colored bathrooms. And that was that. No one ever said another word to her about it.

She also made the decision to bring a bag lunch and eat at her desk, something many of the employees did. Why should she spend the extra money on lunch? It was more convenient as well; the cafeteria was just far enough from her building to have to drive, and who wanted to do that? And it was healthier too, what with the temptation of the ice cream that the cafeteria sold for dessert. Of course, for Katherine Goble, eating at her desk also had the benefit of removing the segregated cafeteria from her daily routine, another reminder of the caste system that would have circumscribed her movements and thoughts. Those unevolved, backward rules were the flies in the Langley buttermilk. So she simply determined to pluck them out, willing into existence a work environment that conformed to her sense of herself and her place in the world.

As the months passed, Katherine stretched out into the office, as at ease as if she had never been anyplace else. Erma Tynes, the other black computer who had been assigned with Katherine, was "by the book": at her desk and working at 7:59:59, barely removing her eyes from the task at hand until the end of the day at 4:30. Katherine, on the other hand, like the engineers around her, got into the habit of reading newspapers and magazines for the first few minutes of the day. She perused *Aviation Week*, trying to connect the dots between the latest industry advance and the torrent of numbers flowing through her calculating machine.

Katherine's confidence and the bright flame of her mind were irresistible to the guys in the Flight Research Division. There was nothing they liked more than brains, and they could see that Katherine Goble had them in abundance. As much as anything, they responded to her exuberance for the work. They loved their jobs, and they saw their own absorption reflected back at them in Katherine's questions and her interest that went so far beyond just running the numbers.

With her fair skin and dulcet West Virginia accent, Katherine might have occupied a fluid racial middle ground, easing her acceptance at the center. Even some of the black employees weren't always sure upon meeting her if she was black. On one occasion, when her mother was visiting from West Virginia, she'd had to take her to the hospital. After an unusually long wait, a doctor had to step in to get her mother put into a room: the admitting desk was moving slowly because they couldn't figure out if she should have a black or a white roommate. One time Katherine's boss, Al Schy, was asked if his group had any black mathematicians. Even with Katherine sitting within earshot, he'd had to think before coming up with a yes. To her colleagues, she had become simply "Katherine."

For any number of reasons, concrete and ineffable, there was something about Katherine Goble that made her as comfortable in the office in 1244 as she was in the choir loft at Carver Presbyterian. She didn't close her eyes to the racism that existed; she knew just as well as any other black person the tax levied upon them because of their color. But she didn't feel it in the same way. She wished it away, willed it out of existence inasmuch as her daily life was concerned. She had taken the long road to Langley's Flight Research Division, but she knew with a confidence approaching 100 percent that she had arrived at the right destination.

"I want to move our girls out of the projects," Jimmy Goble said to Katherine after two years in Newport News.

Moving to Newsome Park had made it possible for Katherine and her family to adapt quickly to life in Hampton Roads. The neighborhood, with its ties to the shipyard and to Langley, with residents who were connected to virtually every aspect of black life in the region, had provided them and their family with a ready-made community. In defiance of the newspaper headlines, Newsome Park had managed to persevere against the ever-present specter of demolition: with the flare-up of military tensions in Korea in 1950, the federal Housing and Home Finance Agency again decided that Newsome Park and all similar housing projects were necessary to the country's ongoing defense effort. The residents of the neighborhood heaved a collective sigh of relief.

More than matters of international law at the 38th parallel, which divided Russian-allied North Korea from US-friendly South Korea, it was the local law of supply and demand that was really keeping Newsome Park off the chopping block. Years after the end of the war, the shortage of adequate housing for the area's Negro residents was still the reality. If the government decided to demolish Newsome Park tomorrow, there simply would be no place for the residents to go.

But the number of houses in smaller neighborhoods had continued to increase, drawing the attention of upwardly mobile families who, like their white counterparts, had a vision of postwar success that included home ownership. Gayle Street, a cul-de-sac not far from the Buckroe section of town, was an attractive new neighborhood where Chubby Peddrew and her husband bought a house. Aberdeen Gardens, the sprawling development built on former Hampton Institute farmland, was another desirable location, its wide streets with grassy medians and surrounding forests drawing many active-duty and retired military families.

Katherine and Jimmy decided to buy a lot in Mimosa Crescent, the World War II–era neighborhood in Hampton that had been built for middle-class families. The developers of the subdivision had jumped every hurdle the Federal Housing Administration could throw at them, making assurances as to the quality of the neighborhood's homeowners and even putting in place restrictive covenants so that the buyers would not be disqualified from receiving federally insured bank loans, as was the case for many—perhaps most—black neighborhoods around the country. Thomas Villa, one development in Hampton that could not secure financing from local banks, pointed its buyers to the North Carolina Mutual Life Insurance Company, at the time the largest black-owned business in the United States, for home loans.

In 1946, Mimosa Crescent had expanded from its original twenty-two parcels to a total of fifty-one, slowly but surely over the next ten years attracting families who filled in the empty lots with comfortable three- and four-bedroom brick houses. What a thrill, not just to imagine a dream house but to plan the color of the tile in its bathrooms, the wood of the cabinets in the kitchen, the size of the floorboards in the living room! Joylette, as the eldest daughter, would even get her own bedroom, the kind of luxury that most girls—of any color—only got to see in the movies or between the pages of a Nancy Drew mystery. The subdivision's proud residents seeded their lawns and planted trees for shade, hosting patio parties and many a club meeting at their homesteads. The Goble family was soon to join them.

It was the perfect plan . . . until over the course of 1955, Jimmy started to feel sick, first with headaches that kept getting worse, then weakness. But unlike the undulant fever that had afflicted him more than a decade prior, he didn't get better. It took months for doctors to diagnose his ailment. They finally discovered a tumor, awkwardly located at the base of his skull, and declared it inoperable. He took to his bed, eventually so enervated that he was forced to take an indefinite leave from his job at the shipyard. His health declined slowly but inexorably for more than a year, much of that time spent in the hospital. Katherine and her daughters visited him as often as they could at his sick bed, holding vigil for the most important man in their lives.

James Francis Goble died on a Thursday, just five days before Christmas 1956. Three days later, Carver Memorial Presbyterian Church filled with mourners, the community offering its condolences and support to the young widow and her three adolescent daughters. Joylette, Kathy, and Connie would never again be able to experience the joy of the holiday season without also reliving the ache that came from their father's death. Both Jimmy's parents and Katherine's parents stayed in town through the end of the year. Katherine's in-laws and their families, particularly the Eppses and the Kanes, who lived in Newport News, shared the burden of grief. Jimmy's Alpha Phi Alpha fraternity brothers and Katherine's Alpha Kappa Alpha sorority sisters kept watch over them, bringing food and running errands and taking care of the mundane necessities that seemed impossible to deal with in the face of such a profound loss.

The Goble girls were as devoted to their father as Katherine was to hers. The loss of Jimmy's protective arms and ready smile, the instability that came from the abrupt, premature end of the partnership between their parents, turned their safe world inside out and forced them to shed the fuzzy comforts of childhood for the hard realities of the adult world.

But Katherine would not yield to loss and chaos. She had made a solemn promise to her husband that she would do everything in her power to keep their bright, lively daughters on the path they had so carefully paved for them from the beginning of their lives. Katherine allowed herself and her girls until the end of the year to give themselves over completely to bereavement. On the first day of school in January 1957, following the Christmas holidays, she accompanied her daughters to a meeting with the school principal. "It is very important that you don't show the girls any special treatment, or let up on them in any way," Katherine said to the principal. "They are going to college, and they need to be prepared." With her daughters, she established the new rules of a household run by a single mother: "You will have my clothes ironed and ready in the morning, and dinner ready when I come home," she instructed. Katherine was now the mother and father, the love and the discipline, the carrot and the stick, and the sole breadwinner.

Katherine and Jimmy shared great ambitions for their children. The Goble sisters excelled in school and took piano and violin lessons and practiced diligently. They were good-natured, outgoing, and respectful, and they always rose to the high standards their parents had set for them. In her children, Katherine saw the legacy of her parents and Jimmy's parents and all their generations past, each pushing their energy and resources to the limits to lift their progeny toward the American dream, to a life that would surpass their own in material and emotional richness and access to the long-promised blessings of democracy. Everything depended on Katherine's ability to hold her family together; she could not fall apart. Or perhaps she *would* not fall apart. There was, and always had been, about Katherine Goble a certain gravity, a preternatural self-possession that had made it the most logical thing in the world that she would teach Roman numerals to the president's brother or converse in French with visiting aristocrats. She seemed to absorb the short-term oscillations of life without being dislodged by them, as though she were actually standing back observing that both travail and elation were merely part of a much larger, much smoother curve.

Certainly much of Katherine's equipoise came from her father, Joshua. Family lore had it that he possessed unexplained skills and senses, that his nimble hands could spirit away afflictions in both humans and animals. Even after he went to work for the Greenbrier, neighbors black and white would call on him to see sick horses through a period of crisis. Years later, Joshua Coleman's granddaughters would recall their grandfather saying that from their first meeting, he had a premonition that Jimmy Goble would not live a long life. Perhaps Katherine, with some intuition of her father's vision, drew strength from the knowledge that her husband's premature death was part of a way of things, however painful.

Or maybe it was her father's pragmatic dictum—"You are no better than anyone else, and no one is better than you"—that disposed her to see the hardships of her life as a fate shared by everyone, her good fortunes as an unearned blessing. With her father's words to buoy her, Katherine Goble observed the manifestations of segregation at Langley, decried the injustice they represented, yet did not feel their weight on her own shoulders. Once she crossed the threshold of Building 1244, she entered a world of equals, and she refused to behave in any way that would contradict that belief.

It was a part of her nature that some of the other black employees at Langley found mysterious, even vexing. How could she be so dismissive of the racism in their workplace, however passive, when her very entry to the laboratory had been under segregated circumstances? Katherine Goble's genuine comfort with the white men she worked with allowed her to be herself with them, no mask required. When the Supreme Court announced the *Brown v. Board of Education* verdict ending legalized school segregation in 1954, she and the engineers had a long conversation about it, talking about the matter forthrightly rather than avoiding it the way a driver swerves to keep from hitting a fallen tree in the road. ("We decided we were all for it," she remembered.) Perhaps as much as Katherine's expectation that she should be treated as the equal of the engineers she worked with was her willingness to treat *them* as equals—to acknowledge that their intellect and curiosity matched hers, that they were bringing to the professional relationship the same sense of fairness and respect and goodwill that she was—that paved the way for her ultimate success.

Jimmy Goble's death cleaved Katherine's life in two parts. They had walked side by side through graduate school and marriage, children and the move to Newport News. Now, at just thirty-eight years old, she found herself a widow and a mother, but also a professional still in the early days of realizing her long-held dream. Jimmy wouldn't be there to see it come to its fruition, but with love, support, and a belief in her talent, he had escorted her to the threshold, and she would carry his spirit and their memories forward. And so Jimmy Goble's death at the end of 1956 wasn't so much an end as an intermission. All that had come before would connect to all that was to come. In January 1957, Katherine's daughters went back to school, and she went back to work: the second act of her life was about to begin.

CHAPTER FOURTEEN

Angle of Attack

By the 1950s, Dorothy Vaughan was also looking forward to a time of change, imagining an era when she and the other computers who wore skirts would be forced to concede ground to the inanimate computers that were redefining the technological frontier. As much as any other profession, aeronautical engineering embodied the restlessness and technological progress that characterized what was already being dubbed the American Century. Jet engines were replacing propellers. The fulfillment of Mach 1 fed the appetite for Mach 2. Supersonic begat hypersonic. Curiosity would not be sated until the mechanical finches that were now so abundant around the globe had evolved to fly to the limits of the atmosphere.

With the complexity that attended the relentless advance of aeronautical research came the need for a new machine. In 1947, the laboratory bought an "electronic calculator" from Bell Telephone Laboratories, an investment in the ongoing need for transonic flight research. Modeling flight at transonic speeds was a particularly knotty problem, because of the subsonic and supersonic winds that passed over the plane or model simultaneously. Aerodynamic equations describing transonic airflows might contain as many as thirty-five variables. Because each point in the airflow was dependent on the others, an error made in one part of the series would cause an error in all the others. Calculating the pressure distribution over a particular airfoil at a transonic speed could easily take a month to complete for the most experienced of mathematicians. The Bell calculator accomplished the same task in a few hours.

No one would confuse the women who used mechanical calculators to process research data with the room-sized electronic devices that performed the same function. Langley put a former East Computer named Sara Bullock in charge of a group dedicated to using the squat, gunmetal-gray block to solve engineers' equations. Already considered superior to the University of Pennsylvania's pioneering ENIAC computer, the Bell used paper punch tapes as input and chugged along at two seconds per operation. The whole building shook when it ran, but it generated answers sixteen times faster than the human computers, with the additional benefit that after the women went home for the day, the Bell could be left running overnight.

In the mid-1950s the center bought its first IBM computers—an IBM 604 Electronic Calculating Punch, then an IBM 650. Originally destined for the lab's finance department, enterprising researchers soon appropriated the machines for their own purposes. One of the uses was to calculate a trajectory—a detailed flight path—for a hypersonic "rocket plane" called the X-15, an experimental vehicle engineered to fly high and fast enough to leave Earth's atmosphere and reach the threshold of what was considered "space."

The early data-processing machines weren't paragons of reliability. They made mistakes, and engineers—or, more likely, the human (female) computers who worked for them—had to keep a close watch on machines' output. "That's not right!" "Let's run it again!" engineers would say to the machine's operators, just like John Becker had said to Mary Jackson. But even with the errors, the machines processed the transonic, supersonic, and hypersonic flows and trajectory analyses in a way that surpassed the upper limits of human ability. In the 1950s, most of Langley's test data was still processed by hand; the NACA's entire research operation had evolved with the women's work as its engine. Electronic computers were rare jewels, their million-dollar-plus price tags affordable only by large research universities and government institutions. And for all their advantage in speed, the computers could still process only one job at a time. The devices chug-chug-chugged around the clock, but competition for computing time on the machines was fierce.

Only the most shortsighted, however, failed to recognize that electronic computers were around for the long haul. Electronic data processing machines brought otherwise unattainable power and efficiency to the research process. There was no reason to think that they wouldn't continue to poach more tasks that were currently completed by hand. Evolution occurred in scientific progress as it happened in nature: a positive trait was passed along, then proliferated; obsolete characteristics withered away, and the technology and the organization evolved into something new. Propeller research, for example, had been one of Langley's most important lines of inquiry from its inception through the lead-up to the war. By 1951, the Propeller Research Tunnel was declared obsolete and demolished, and the engineers that had staffed it were left to find a new specialty or retire.

The female mathematicians' job security wasn't immediately threatened by the machines, but Dorothy Vaughan perceived that mastering the machine would be the key to long-term career stability. When Langley sponsored a series of computation courses to be held after work and on weekends, Dorothy wasted no time enrolling. She encouraged the women in her group to do the same.

"Integration is going to come," she told her employees. The blurring of the color lines could put her and her reports in a position to qualify for the desirable jobs that were sure to open for people who were experts in managing the electronic computers. To keep moving forward, they needed to take advantage of every opportunity to make themselves as valuable as possible to the laboratory.

Scientific progress in the twentieth century had been relatively linear; social progress, on the other hand, did not always move in a straight line, as the descent from the hopeful years after the Civil War into the despairing circumstances of the Jim Crow laws proved. But since World War II, one brick after another had been pried from the walls of segregation. The Supreme Court victories opening graduate education to black students, the executive orders integrating the federal government and the military, the victory, both real and symbolic, when the Brooklyn Dodgers signed Negro baseball player Jackie Robinson, were all new landings reached, new corners turned, hopes that pushed Negroes to redouble their efforts to sever the link between separate and equal decisively and permanently.

Farmville, the town that Dorothy left behind in the 1940s, had become in the 1950s a microcosm of America's struggle over integration in its public schools. In the thirteen years since she'd left Moton High School, the deficient building had passed from being merely over-crowded to packed beyond reasonable measure. In 1947 the state constructed tar paper shacks on the school's lawn (the students called them "chicken coops") in an attempt to squeeze 450 students into a school built for 180. In 1951, one of the school's decrepit school buses crashed, killing five students. One of the victims was the best friend of Barbara Johns, the sixteen-year-old niece of Farmville native and renowned civil rights activist Vernon Johns, who, at the time of the accident, was a preacher at a church in Montgomery, Alabama.

The grief that washed over Barbara Johns gave way to anger, then took hold in her as a hunger for justice that would not be denied. In April 1951, the same month Langley promoted Dorothy Vaughan to the head of West Area Computing, Barbara Johns organized her fellow Moton High School students in a walkout, imploring them to take a public stand against the abysmal conditions at the school; she stood strong, leading the charge through the opposition and fear of many parents and teachers. Dorothy's nieces and nephews were among the strikers. At the time, none of them could have foreseen the consequences of the dominoes that the coura-geous teenagers set in motion in 1951: Barbara Johns' campaign to attend a school that equaled the standards of white Farmville High attracted the attention of Virginia lawyers Spottswood Robinson and Oliver Hill, who then joined forces with Thurgood Marshall, the NAACP's chief counsel. Marshall consolidated the Moton students' suit with four others around the country into the US Supreme Court case *Brown v. Board of Education*, the landmark 1954 decision that banned segregation in all public schools in the United States of America. Black Americans sent up a cheer of jubilation, and the ruling provided momentum and hope to grassroots civil resistance and social movements throughout the land. "Not Willing to Wait: NAACP Leaders Want Integration 'Now!' " declared a *Norfolk Journal and Guide* headline.

Waiting, though, was exactly what Virginia's leading politicians, starting with Senator Harry Byrd, had in mind. "If we can organize the Southern States for massive resistance to this order I think that, in time, the rest of the country will realize that racial integration is not going to be accepted in the South," Byrd said in the wake of the Supreme Court decision. Virginia's resistance to the ruling would, over time, be more intransigent and longer lasting than that of any other state. When Dorothy and the other West Computers signed up for computation classes in the 1950s, they registered to attend at Hampton Institute. Langley offered on its premises a series of lectures on aerodynamics, open to all comers. It held one engineering course on-site, which some of the black employees attended. It had set up a classroom on the air force base, a cooperative venture with George Washington University, presumably available to all employees. Nearby College of William and Mary extended its classrooms to Langley employees. Newport News High School held night classes. Langley managed so many courses in so many places that it often seemed like a university itself.

Hampton High School was the seat of the University of Virginia's Extension School, and the most significant of Langley's campuses. In the evenings, the city's only public high school taught laboratory employees everything from sewing to dynamic model design, bookkeeping to machine shop theory. It even hosted an Americanization class, helping foreign employees prepare for the citizenship test. Most of the classes covered math, science, and engineering. The lineup included courses like Differential Equations, a core part of the engineering curriculum, and higher-level math, such as Theory of Equations.

But the high school was off-limits to the city's Negro children, who were still sent to Phenix High School, Mary Jackson's alma mater. In 1953, a Negro lawyer named William Davis Butts had gone before the Hampton school board to decry Phenix's "inadequate gymnasium and library" and to demand that the city "terminate the 'undemocratic and expensive dual system.' " The board, deferring to the state segregation law, declared his pleading moot. As Hampton's schools remained segregated for its schoolchildren, the UVa Extension Program rebuffed Langley's black employees. More than a decade after the first West Computers headed to Hampton Institute for wartime ESMWT classes, Langley's black professionals still relied on the august black college for professional training and career advancement.

Across the country, the United States debated the quality of its schools, concerned with how American students matched up to the Soviets in math and the sciences. The imperative to raise the general level of technical proficiency had only grown stronger as the relationship between the United States and the Soviet Union grew more inflamed. While the discussion in World War II had centered on using white women in engineering and science, the 1950s debate had expanded to a broad discussion of the participation of Negroes in the technical fields as well. Virtually every review of the situation questioned how much desperately needed brainpower was being squandered by the intentional neglect of America's Negro schools.

Kaz Czarnecki wasn't about to leave brainpower on the table. He only learned of Mary Jackson's double major in math and science after he made her the offer to join the Four-Foot SPT group. Even so, without having reviewed her résumé, something about her gave him the idea that she was both qualified and the right fit for the job. He was white, male, Catholic, and a Yankee. She was a black woman from the South, a devout member of the African Methodist Episcopal Church. It would have been easy for each of them to look past the other, to see the outside and assume that they could have nothing in common. But what Kaz

Czarnecki intuited, and what the years would bear out, was this: Mary Jackson had the soul of an engineer.

From the beginning, Czarnecki had put Mary at the controls in the wind tunnel, showing her how to fire up the tunnel's roaring sixty-thousand-horsepower engines (the noise from years of work in the tunnel eventually damaging Mary's hearing). He showed her how to work with the mechanics to correctly position a model in the test section. One test required Mary to clamber onto the catwalk of the wind tunnel, measuring how rivets disrupted the airflow over a particular model. Another involved turning the tunnel's Mach 2 winds on a series of sharp-nosed metal cones to discover the point at which the smooth air flowing over the cones became turbulent. The research had application to the design of missiles, of great interest as the United States sought to gain every possible military and technological advantage over the Soviet Union. The results of the work would come to fruition in 1958, in Mary's first report, coauthored with Czarnecki: "Effects on Nose Angle and Mach Number on Transition on Cones at Supersonic Speeds," published in September 1958.

It wasn't long before Mary's new boss suggested that she enroll in the laboratory's engineer training program; her ability and her passion for the work were abundantly clear. Most important, she now had a sponsor, a mentor willing to make her career and prospects for advancement his responsibility. The majority of Langley's female professionals had spent their time at the laboratory classified as computers. Some, like Dorothy Vaughan and Dorothy Hoover, made the grade as mathematicians from day one; others earned the designation over time. In the mid-1950s, a woman named Helen Willey led a successful charge to have every female computer with a math degree upgraded to mathematician, a title that automatically applied to men with the same credential. Regardless of this gain, nearly all the women still worked at the behest of an engineer. It was the engineer who determined what problems to investigate, designed the experiments, and defined the assignments for the mathematicians. Engineers gave direction to the craftsmen who made the wind tunnel models and to the technicians and mechanics who manipulated the models. It was the engineer who faced the firing-squad editorial review board to defend the collective effort represented by the research report, and it was the engineer who took the victory lap when the report was published.

Most of the country's top engineering schools didn't accept women. Kitty O'Brien Joyner, the laboratory's first female engineer from the time Pearl Young left until the middle of the 1950s, had been forced to sue the University of Virginia to enroll in the school's all-male undergraduate engineering school in 1939. As for black female engineers, there weren't enough of them in the country to constitute a rounding error. In 1952, Howard University had had only two female engineering graduates in its history. Being an engineer, Mary Jackson would eventually learn, meant being the only black person, or the only woman, or both, at industry conferences for years. Kaz's endorsement put Mary on the engineering track, essentially promising a promotion when she successfully completed a few core courses. For Mary, differential equations were the first step. Actually, it was not that simple. The first step was to get permission to enter Hampton High School. If Mary had applied for a job as janitor, the doors to the school would swing wide open. As a professional engineer-in-training with a plan to occupy the building for the nefarious purpose of advancing her education, she needed to petition the city of Hampton for "special permission" to attend classes in the whites-only school.

Mary was seeking to make herself more useful to her country, and yet it was she who had to go hat in hand to the school board. It was a grit-your-teeth, close-your-eyes, take-a-deep breath kind of indignity. However, there was never any doubt in Mary's mind that it must be done. She would let nothing—not even the state of Virginia's segregation policy—stand in the way of her pursuit of the career that had rather unexpectedly presented itself to her. She had worked too hard, her parents had worked too hard; a love of education and a belief that their country would eventually heed the better angels of its nature was one of their great bequests to their eleven children.

The City of Hampton granted Mary the dispensation. The pass gave her access to the classes, though it did not make them broadly available to others. Whatever pain securing the permit exacted, it was more than offset by the victories lying in wait. She began her coursework at Hampton High School in the spring of 1956.

Mary Jackson had passed by the old Hampton High School building too many times to count. The local landmark was located in the middle of the city, not far from her home downtown. Her night school classmates were the same daytime colleagues she had known for five years, but it was only natural that she should be anxious at the thought of meeting them on the other side of the physical, emotional—and legal—threshold she was about to cross. Nothing, however, could have prepared her for the shock that awaited her when she walked through the long-closed door.

Hampton High School was a dilapidated, musty old building.

A stunned Mary Jackson wondered: was this what she and the rest of the black children in the city had been denied all these years? This rundown, antiquated place? She had just assumed that if whites had worked so hard to deny her admission to the school, it must have been a wonderland. But this? Why not combine the resources to build a beautiful school for both black and white students? Throughout the South, municipalities maintained two parallel inefficient school systems, which gave the short end of the stick to the poorest whites as well as blacks. The cruelty of racial prejudice was so often accompanied by absurdity, a tangle of arbitrary rules and distinctions that subverted the shared interests of people who had been taught to see themselves as irreconcilably different.

It was the kind of thing Mary would shake her head about, laughing to keep from crying, with Thomas Byrdsong, a black engineer who had come to Langley in 1952. Byrdsong was a Newport News native who served in World War II in the Montford Point Marines, the first group of black men permitted to join that previously restricted branch of the American military. Another University of Michigan engineering grad who followed Jim Williams's path to Langley, Thomas Byrdsong was a frequent guest at Mary and Levi Jackson's welcoming dinner table, always pleased to sample Levi Jackson's delicious home cooking and enjoy the warmth of an evening unwinding with the down-to-earth couple. There, they could talk Reynolds numbers and aeronautical shop and let their guard down about the challenges of their jobs. Being on the leading edge of integration was not for the faint of heart.

Fresh out of the University of Michigan, Thomas Byrdsong had been assigned to Gerald Rainey, a senior engineer in the Sixteen-Foot Transonic Dynamics Tunnel. Rainey instructed Byrdsong in the procedures for conducting his first test in the tunnel, assigning an experienced mechanic to assist his wet-behind-the-ears engineer. The mechanic, a white man with many years of service at the laboratory, sabotaged Byrdsong's experiment by incorrectly affixing the model to its sting in the tunnel's test section. The problem, and the cause of it, was obvious to Rainey as soon as he sat down with Byrdsong to review the test data, which had been contaminated by the mechanic's vicious prank. Rainey upbraided the mechanic in Thomas Byrdsong's presence. "You will never do that again to this man or anyone else, do you understand me?" Rainey shouted at the mechanic.

As a son of the South, Thomas Byrdsong knew all too well the consequences that might befall a black man who openly expressed his anger in front of white people. He went out of his way to maintain a calm demeanor at work, but internalized anger came at a cost, and he took to frequenting the bar at the local Holiday Inn after work—one of the few integrated public places in the city—for a little liquid attitude adjustment before going home to his family.

In general, the black men at Langley—in 1955, Lawrence Brown joined Jim Williams and Thomas Byrdsong—were more likely to run into the minefield of race than the women. Their impeccable manners and graciousness did not offer complete protection from the reactions some of the staff had to the presence of black men in professional positions at the laboratory. Most white engineers were cordial with the black men, even anxious to protect them against racist incidents, as Rainey had Byrdsong. It was from the blue-collar mechanics, model makers, and technicians, many of whom hailed from local "sundown towns" such as Poquoson, where blacks were not welcome, that they usually caught hell.

Tall, brown-skinned, and unmistakably black, there would be no tiptoeing into the white bathrooms for Jim Williams and Thomas Byrdsong. Like Katherine Goble, however, they also figured a way to opt out of the segregated facilities. Each day at lunchtime they escaped to a black-owned restaurant just outside the entrance to the air force base for relief and a little home cooking, circumventing both the cafeteria and the Colored men's bathroom at Langley.

The events of the next few years would test the United States on every front: on secret battlefields in faraway countries, in the classrooms and voting booths of the South, in the halls of Congress and on the streets of Washington, DC. The competition between the United States and the Soviet Union for control of the heavens and the earth was about to escalate in a way that would push the steely intellect of every single one of the brain busters on the NACA rolls to its limits. Each upheaval caused Americans of every background to ask themselves and each other, What are we fighting for? Black Americans knew, and they answered as they had each time their country called: for democracy abroad *and* at home. So they took up arms again: on the battlefields, in the classrooms and voting booths, in the nation's capital—and in the offices of the Langley Aeronautical Laboratory.

CHAPTER FIFTEEN

Young, Gifted, and Black

October 5, 1957, was the kind of day that never failed to delight Christine Mann, a rising senior at the Allen School for girls in Asheville, North Carolina. While the rest of her boarding school classmates still clung to the last precious moments of sleep, Christine left her dorm and headed to her daily job at the library, setting out the newspapers and magazines that the school received each day. As she walked across the campus, the world moved from shadow to sun, the purple-blue peaks that stood watch over the town shrugging off the mist that lent the Great Smoky Mountains their name. The light of day revealed fall's early brilliance, chartreuse, gold, and orange leaves eclipsing the dark green of summer. The scarlet of the maple trees appeared in just a few brilliant spots; in a month's time, the red would spread like a fever, dominating the landscape.

Christine collected the periodicals left in the newspaper box and unlocked the library door. Her daily job of setting out the newspapers was simple but came with responsibility, as it meant that the faculty entrusted her with the keys to the library. The quiet time she spent alone there, a modest brick building filled with walnut furniture and fragrant with the must of old books, was the best part of her job. Every morning before the library filled with students, she perused the newspapers, taking in the events of the previous day.

Since the beginning of the school year, newspapers around the country had bled with coverage of the crisis in Little Rock, Arkansas. Nine black teenagers trying to integrate all-white Central High School had turned the state's capital city into a military battleground. At the command of the governor, Orval Faubus, the Arkansas National Guard had been called out to prevent the black students from entering the school. Three days later President Eisenhower trumped the state, federalizing the state's guard and sending in US Army troops to escort the nine through the school's doors. The crisis unfolded over days, each morning a new installment, and always accompanied by photos that were as difficult to look at as they were to look away from: images of black students Christine's age, their arms heavy with books, struggling to maintain composure as a phalanx of military men protected them from the screaming, spitting, bottle-throwing white crowd that surrounded them. All for wanting the key to what Central High School and the all-white schools across the South had kept locked away from black students like her. Christine allowed herself to walk in their shoes for a moment, wondering how she would deal with the taunts, the bottles, the epithets, and the humiliation. It would come as a relief to finish the article and find herself back in the haven of Allen's library.

As Christine read the Little Rock coverage, so did the rest of the country—and the world. In Europe, and in the capitals of Asia and Africa, people devoured the particulars of the Little Rock crisis. Photos of the black students being threatened with violence for the pursuit of education, along with the details of lynchings, subjugations, and other injustices issuing forth from the South, undermined the United States' standing in the postwar competition for allies. No matter how hard the United States tried, despite the best efforts of its diplomatic corps and its propaganda machine, it seemed impossible to divert the eyes of the world from the ugliness unfolding in Little Rock and all of its implications for the legitimacy of American democracy. That is, until a Soviet gambit changed the conversation.

"Red-Made Satellite Flashes Across US," printed the *Daily Press* in Newport News. "Sphere Tracked in 4 Crossings Over US," ran the *New York Times* headline. It took no time for the mysterious name to pass from Soviet mouths to American ears: Sputnik. Radio Moscow announced a timetable, revealing exactly where the satellite would fly over Earth, and when. Christine had fallen asleep in one world and awakened in another. October 4, 1957, was the midnight of the postwar era, and the end of the naive hope that the conflict that was ended by an atomic bomb would give way to an era of global peace. The morning of October 5 was the official dawn of the space age, the public debut of man's competition to break free of the bonds of terrestrial gravity and travel, along with all his belligerent tendencies, beyond Earth's atmosphere.

That morning of October 5, absorbing the impact of the initial headlines, Christine experienced a mix of emotions. Fear, certainly: she was just three years old when a B-29 Superfortress dropped an atomic bomb on Japan, forever linking the name Hiroshima with annihilation. She and her generation were the first in the history of the world to come of age with the possibility of human extinction as a by-product of human ingenuity. As hostility between the United States and the Soviet Union increased, it began to feel like a probability. Black-and-yellow triangular fallout shelter signs proliferated in public spaces, pointing the way to underground refuge from radiation. Christine dutifully submitted to civil defense drills at school, ducking and covering under her desk, practicing the maneuver that adults said would protect her and her classmates from that telltale flash "brighter than the sun."

While students and teachers hoped their desks and basements would stand up to the power of a nuclear blast, the country's leaders also prepared for a possible attack—in high style. In one of the Cold War's most unbelievable episodes, in 1959 President Eisenhower authorized the construction of a secret bunker deep under the Greenbrier hotel, the resort in White Sulphur Springs, West Virginia, where Katherine Goble, her father, Joshua Coleman, and Dorothy Vaughan's husband, Howard, had all worked. Dubbed "Project Greek Island," in the event of an attack on Washington, DC, senators and congressional representatives were to be evacuated from the nation's capital by railroad and delivered to the Greenbrier's bunker. There was no room in the bunker for spouses or children, but it was stocked with champagne and steaks for the politicians themselves. The luxury underground fortress remained operational and ready to receive its political guests until a 1992 exposé by *Washington Post* reporter Ted Gup blew the operation's cover.

Initially, President Eisenhower tried to pooh-pooh the Russians' "small ball in the air" as an insignificant achievement, but the American people would have none of it. Sputnik, some experts declared, was nothing less than a technological Pearl Harbor.

For the third time in the century, the United States found itself trailing technologically during a period of rising international tension. On the cusp of World War I, the country's inadequate supply of aircraft had given birth to the NACA. The mediocre American aircraft industry of the 1930s rose to preeminence because of the challenge of World War II. What would it take for the country to prevail against this latest threat? Sputnik was proof, American policymakers assumed, that the Soviet Union had intercontinental ballistic missiles—many of them, hundreds perhaps, with the power to hurl an atomic weapon at US cities. A new term began to make the rounds in policy circles, the press, and private conversation: the missile gap.

Black newspapers and their readers wasted no time in making the link between America's inadequacy in space and the dreadful conditions facing many black students in the South. "While we were forming mobs to drive an Autherine Lucy [the black woman who integrated the University of Alabama in 1956] from an Alabama campus, the Russians were compelling ALL children to attend the best possible schools," opined the *Chicago Defender*. Until the United States cured its "Mississippiitis"—that disease of segregation, violence, and oppression that plagued America like a chronic bout of consumption—the paper declared, it would never merit the position of world leadership. An editorial in the Cleveland *Call and Post* echoed that sentiment. "Who can say that it was not the institution of the Jim Crow School that has deprived this nation of the black scientist who might have solved the technological kinks delaying our satellite launching?" wrote the paper's editor and publisher, Charles H. Loeb.

But segregation couldn't restrain Christine's curiosity. Along with the anxiety that the Russians' accomplishment provoked, Christine felt a sense of wonder, even a thrill, to see the skies above open so wide. The world beyond Earth had always been a mysterious place, silent, dark, and cold, the realm of magic and gods. Wernher von Braun, the former Nazi rocket scientist granted amnesty by the United States after World War II in exchange for helping the country build a dominant missile program, functioned as the nation's head space cheerleader. A series of articles that von Braun contributed to *Collier's* magazine in 1952—"Man Will Conquer Space Soon!"—presented space travel as the natural next step for the restless inhabitants of the Earth. American television viewers tuned in religiously to science fiction television programs like *Space Patrol* and *Tales of Tomorrow*. But Sputnik was anything but fiction, and it was happening *today*.

Christine also took umbrage at the Soviets' excursion into the heavens. From her core came the desire to rise up to meet the gauntlet they had thrown down. She was an American, after all, and the Russians were the enemy! We can't let them beat us, she thought, echoing the sentiments of virtually every American citizen. It would take time for her to work it out, but somehow, even in those first moments of learning of the Soviet accomplishment, she believed that this was her fight too.

The Soviet Union also thought it was Christine's fight. Four days after launching Sputnik into orbit, Radio Moscow announced the addition of one more city to its timetable of destinations that would be overflown by their satellite: Little Rock, Arkansas.

Three years earlier, before Christine's parents had enrolled her as a student at the Allen School, another headline-making event had intruded upon her daily life. On May 17, 1954, she was still enrolled at the Winchester Avenue School in Monroe, North Carolina, her hometown. The principal of the school stepped into her eighth-grade classroom, interrupting the lesson with an announcement. "I just came to let you all know that the Supreme Court just ruled on *Brown v. Board of Education*, and you will be going to school with white students in the future," he said. The same report that sparked conversation among Katherine Goble and her colleagues left Christine and her classmates agape.

Located twenty-five miles down a winding road from Charlotte, Monroe, population seven thousand, was typical small-town South. Everyone in the Newtown neighborhood, where Christine lived, from the doctor to the street sweeper to the teachers at the Winchester Avenue School, was black. Most of the black men in Monroe earned their living working for the railroad line that ran through the town. Black women held jobs in the Monroe Cotton Mill or as domestic servants. Virtually everyone and everything white in Monroe, including the white school and white residents, such as future US senator Jesse Helms, son of a former fire chief, existed across the dozen or so railroad tracks that sliced through the town like a combine. How will we, thought the students at Winchester—with our rickety desks and dog-eared secondhand textbooks, our poorly equipped to nonexistent science laboratories—how *could* we compete with the white kids from the other side of the tracks?

The principal spoke with such gravity that Christine and her classmates worried they might have to pack up their books and decamp for the other side of town at that very moment. Segregation was the only world they had known. Discrimination was the force that concentrated them in Newtown, that enrolled them in the Winchester school, that sent Christine's parents to be educated at Knoxville College rather than the University of Tennessee. Discrimination they had come to expect, if not accept. But the prospect of integration planted a new fear in the souls of Christine and fellow members of the *Brown v. Board of Ed* generation: that as blacks, they would not be good enough—*smart* enough—to sit next to whites in a classroom and succeed.

Christine's parents, Noah and Desma Mann (unrelated to West Computing's Miriam Mann), were the products of the same Negro institutions and the same values—"education, honesty, hard work, and character"—that formed their contemporary, Dorothy Vaughan. In the early years of their marriage, the Manns traveled around Alabama, Georgia, and North Carolina, moving from one teaching job to another. Desma left teaching to care for what would grow to become a family of five children. Noah Mann, eager to make enough money to cover his household's costs and provide for his children's future, eventually took a more lucrative job as a sales rep based out of the Charlotte office of the North Carolina Mutual Insurance Company, the same successful black-owned company that had underwritten the home loans of black home buyers in Hampton, including in Mimosa Crescent, Katherine Goble's neighborhood.

In 1943, the family settled in Monroe, the seat of Union County, Noah's assigned sales territory. The position afforded the Manns a comfortable living, and they were one of the few black families in town who owned a car, a Pontiac Hydramatic, which Christine's father used to go collect premiums from customers. Every day after work, Noah wheeled the big automobile into the driveway and asked his youngest daughter, "What did you learn today?" Sometimes Christine accompanied him on his rounds. When she was barely old enough to see over the windshield, Noah gave Christine driving lessons on quiet country roads. She loved it when her father taught her the tricks like priming the carburetor that would keep the temperamental machine on the road. Bold and curious, Christine learned to ride a bike by rolling at top speed down one of Monroe's many hills, flying off in one direction like a daredevil at the bottom of the hill while the bicycle went banging off in another. Patching tires and adjusting the bike's brakes with a coat hanger became important parts of her mechanical repertoire. Dolls interested her mainly for what was inside them; her mother would catch her tearing out their stuffing so that she could see what was making them talk.

Younger than her closest sibling by eight years, and nearly thirteen years younger than her oldest brother, Christine's early life revolved around the routines of the grown-up world. Shortly after Christine's birth, Desma Mann returned to teaching. Christine stayed home with a babysitter until she was old enough to accompany her mother each day to her job in a two-room elementary school out in surrounding Union County. Across the street from the school stretched acres of cotton fields, the raw material for Monroe's mill and the source of income for many county residents. The school year followed the picking season. Students sweltered in the desks throughout the North Carolina summer before being released in time for the harvest in September and

October. With all potential playmates either in school or working in the fields, Christine entertained herself by joining in the lessons in her mother's classroom. By the time she turned five, Desma Mann's youngest daughter was a second-grade student, ready to attend the consolidated Winchester Avenue School in Monroe.

Christine became best friends with the principal's daughter, Julia. The two were inseparable and went everywhere together. "Julia's parents said she could go. Can I go too?" was Christine's constant query to her parents. But with the onset of adolescence, as requests turned from afternoons riding bikes to dances and socializing with the kids in her class who were two years older, Christine's parents decided to send their daughter to Allen to eliminate the possibility that she might be distracted from her studies.

The Allen School was founded in 1887 by white United Methodist missionaries, with the goal of providing talented Negro girls from Appalachian North Carolina with the best possible start in life. All the girls had "duty work assignments," like Christine's post at the library, a practical way to teach them responsibility and discipline. Many students came from working-class or poor families; Christine was one of the few at the school who did not receive assistance to cover the costs of tuition and board. Despite the economic circumstances of the student body, Allen was considered one of the best Negro high schools in the country. Parents from as far away as New York sent their children to Allen for its rigorous liberal arts curriculum, its religious teaching, and its insistence on imparting social graces to its students. Band leader Cab Calloway's niece attended in the 1940s. A 1950 graduate named Eunice Waymon had made her way from North Carolina to New York and was already in the process of transforming herself into the singer, pianist, and civil rights activist Nina Simone.

Waves of homesickness washed over Christine in the fall of 1956, her first semester away from home. She phoned her parents every chance she could, begging them to let her return to the familiarity of Monroe. As the months rolled by, though, Christine came to love boarding school life. She opened herself to new friends, the stern but doting Methodist faculty, and the school's routine and rituals. A charismatic eleventh-grade geometry teacher stoked her interest in math, and for the first time, she entertained the idea of a future that took advantage of her knack for numbers and all things analytical.

College, of course, wasn't a matter of if, but where. Most Allen graduates went on to higher education, some to prestigious northern schools like Vassar and Smith. In 1956, the University of North Carolina at Greensboro, Virginia Tucker's alma mater, admitted its first black students, Bettye Tillman and JoAnne Smart. In contrast to its neighbor's militant position on segregation, North Carolina made cautious moves to comply with the *Brown v. Board of Education* ruling. "After careful deliberation, it is my opinion that desegregation is an idea whose hour has arrived," said Benjamin Lee Smith, the superintendent of the Greensboro public school system.

Christine, however, decided to follow in the family tradition of attending a black college, but she had long known that she didn't want to walk a path previously worn by her older sisters and brothers. Two of her siblings had attended Johnson C. Smith, in Charlotte; one had graduated from Tennessee State, and another from Fisk, in Nashville. Two years away from home, away from the shelter of her parents and the model of her older siblings, had given Christine the desire, and the confidence, to strike out on her own.

The summer before her senior year, Christine accompanied her friend Julia's family to attend Julia's older sister's graduation from Hampton Institute. Christine didn't know much about the school; she had barely heard the name, but during her visit, she was taken by the elegant campus and green lawns, the balmy breezes of Hampton Roads in May, and the open expanses of coast and ocean. Hampton's student body ranged from youngsters taking their family's first step onto the ladder of social mobility to the scions of the Talented Tenth. The school's strict environment—mandatory chapel, study halls, evening curfew, and a dress code—were so similar to Allen's that Christine would need no adjustment.

Living in Monroe, Christine had always been someone's little sister. At Hampton, she thought, she would become her own woman. In the fall, she applied to the school, with Fisk as her backup plan. Hampton responded with an offer letter and a scholarship covered by the United Negro College Fund.

"I've been accepted at Hampton," Christine wrote her mother in a letter in early 1958. "I have a scholarship at Hampton, and so there is no reason why you shouldn't let me go." Desma Mann fretted at the idea of her baby going off so far away, all alone, but she had always known that day would come. One by one, she had encouraged her children to leave Monroe. There was nothing for them there—no job, no future. Only by leaving home would her children have the chance to reach the potential she and Noah had worked hard to cultivate in them.

Christine graduated from Allen in May 1958. From the time Sputnik took flight in October 1957 until she addressed her classmates as valedictorian, the Soviets launched two more satellites, Sputnik II, carrying the space dog Laika, and Sputnik III. The United States, playing catch-up, managed to put satellites Explorer I and Vanguard I into orbit, though eight of the eleven Vanguard launches failed. The post-Sputnik lament over the lack of American scientists, engineers, mathematicians, and technologists moved President Eisenhower to initiate the National Defense Education Act, a measure designed to cultivate the intellectual talent required to generate successes—short and longer term—in space.

While "Red engineering schools" in the Soviet Union were "loaded with women"—one-third of Soviet engineering grads were female, the *Washington Post* reported in 1958—the United States still struggled to find a place for women and Negroes in its science workplace, and in society at large. The restlessness that disturbed Christine's home state in the form of student protests in Greensboro would follow her, and engage her, at Hampton Institute. And though it would take years for her to realize that Hampton would be her basic training for the "civilian army of the Cold War," she was just months from meeting some of the successes of an earlier collision between race, gender, science, and war: Dorothy Vaughan's children, Ann and Kenneth; Katherine Goble's daughter, Joylette; and the children of many of the other women who had come to Hampton Roads a generation ago and made it their home.

In August Christine bade adieu to Monroe and drove north with her parents in the Hydramatic, which was more than big enough to accommodate the three of them and the possessions she needed to begin her life at Hampton. The peaks of home flattened out as they approached the coast, and then, like the first time she had come to Hampton, it came into view: the James River. She would never renounce her love for the mountains, but the James—so broad and measured as it joined the Chesapeake Bay, so different from the narrow streams that rushed through the ridges of home—took her breath away. Crossing that river as she closed in on Hampton made her feel like anything was possible.

CHAPTER SIXTEEN

What a Difference a Day Makes

Well into her nineties, Katherine Goble would recall watching the winking dot of light in the sky as vividly as if it were still October 1957. She stood outside in the unseasonably warm autumn nights of that year and tracked the shiny pinpoint as it moved low across the horizon. Around Hampton Roads and throughout America, citizens turned their eyes skyward with a mixture of terror and wonder, eager to know if the 184-pound metal sphere launched into orbit by the Russians could see them as they tried to see it from their backyards. They surfed the radio dial trying to lock on to the artificial moon's beeping, its sound like an otherworldly cricket.

"One can imagine the consternation and admiration that would be felt here if the United States were to discover suddenly that some other nation had already put up a successful satellite." Those words from a letter describing a secret 1946 RAND Corporation proposal to the US Air Force, suggesting that the United States design and launch a "world circling satellite," sounded, in 1957, like the unheeded voice of Dickens' Ghost of Christmas Future. In the 1940s, space research was deemed a little too far out to warrant systematic consideration and development. The Rand report gathered dust.

Now, with Sputnik circling overhead every ninety-eight minutes,

Americans demanded to know how their country, so dominant in its victory in the last war, could have been surprised and usurped by a "backward peasantry" like the USSR. Panic spread from coast to coast: was it possible that the satellite was mapping the United States, with the intent of locking down targets for missile-delivered hydrogen bombs? Fear battled humiliation in the American psyche. "First in space means first, period," declared Senate Majority Leader Lyndon Johnson. "Second in space is second in everything." Could Sputnik signal the end of the country's global political dominance?

In reality, the United States wasn't trailing the Soviet Union quite as badly as it appeared in the wake of the Sputnik crisis. The US Army's Jupiter-C missile had been tested successfully on several occasions, and the Americans were ahead of the Russians in terms of the systems that guided missiles on their trajectories into space. But President Eisenhower had insisted that the nation's first foray into space be presented as a peaceful effort, rather than an explicitly military operation that risked triggering a dangerous retaliation by the Soviet Union. The Americans had planned to launch the first satellite into orbit as part of the International Geophysical Year, a cooperative global science project that ran from July 1957 to December 1958. Physicists, chemists, geologists, astronomers, oceanographers, seismologists, and meteorologists from sixty countries, including the United States and the Soviet Union, collaborated to collect data and conduct earth science experiments, under the mantle of peaceful interchange between East and West. Trumped by Sputnik, the Americans played catch-up. The US Army's Jet Propulsion Laboratory successfully orbited the Explorer I satellite in January 1958. Two months later, Project Vanguard, managed by the US Naval Research Laboratory, also managed to launch a satellite, though the achievement was overshadowed by Vanguard's many rocket failures.

From where Katherine Goble was sitting, upstairs in Langley's hangar, the Soviet move looked rather like a new beginning for the NACA-ites. Skies all over the world bore witness to four decades of successful Langley research, from passenger jets to bombers, transport planes to fighter aircraft. With supersonic military aircraft a reality, and the industry moving forward on commercial supersonic transport, it appeared that the "revolutionary advances for atmospheric aircraft" had run their course. Furthermore, Langley's high-speed flight operations, which had been migrating over the years from the populated Hampton Roads area to isolated Dryden, in the Mojave Desert, were officially ended by a 1958 NACA headquarters edict. As Katherine and her colleagues in the Flight Research Division wondered what was next, Sputnik provided them with the answer.

Space had long been a "dirty word" for the airplane-minded Langley. Congress admonished the brain busters not to waste taxpayer money on "science fiction" and dreams of manned spaceflight. Even in the Langley Technical Library, which was arguably the world's best collection of information on powered flight, engineers were hard-pressed to find books on spaceflight.

That didn't stop Langley engineers from imagining how the missile bodies and rocket engines and reentry problems involved in high-speed flight research might also apply to space vehicles. Any craft that traveled into space first had to traverse the layers of Earth's atmosphere, accelerating through the sound barrier and increasing numbers on the Mach speed dial, before escaping the pull of Earth's gravity and settling into the eighteen-thousand-mile-per-hour speed that locked objects into low Earth orbit, following a circuit of between 134 and 584 miles above the planet. On the return trip, the vehicle skidded through the friction of the increasingly dense atmosphere, building up heat that could reach 3,000 degrees Fahrenheit. NACA engineer Harvey Allen discovered, somewhat counterintuitively, that although the most aerodynamically streamlined shapes were best for slipping out of the atmosphere, a blunt body that increased rather than decreased air resistance was best for dissipating the extreme temperatures on the way back down.

With the US government desperate to gain a foothold in the space race, Langley now could open its garage door and display its wares for the world to see. A group that included Mary Jackson's division chief, John Becker, advocated for a vehicle that was capable of reaching orbital speeds and then gliding back down to Earth like a traditional aircraft, an advanced version of the X-15 rocket plane. It would be an elegant solution to the problem of space, they thought, one that made the hearts of the NACA's old-school "wing men" beat faster.

But the urgency of the competition with the Soviets created pressure to adopt the quickest, surest way into space, even if it was a little crude, or sacrificed long-term spacefaring viability for short-term earthly victory. In the Flight Research Division, Katherine Goble spent her days with her mind and her data sheets full of the specifications of real planes—not plane parts, not model planes, not disembodied wings in wind tunnels but actual vehicles that hurtled humans through the atmosphere. Flight Research's cousins, a "notoriously freethinking" group of engineers called the Pilotless Aircraft Research Division—PARD—had developed an expertise in rocketry, setting up an adjunct operation on an isolated test range on Wallops Island off the Virginia coast. Their rockets had reached speeds of Mach 15 in flight, and they were confident of their abilities to lift a payload—a satellite and a human passenger—into orbit.

As the clamor for action in space grew louder, engineers from PARD and the Flight Research Division moved to center stage. The core of the group coalescing around the US space effort shared an office with Katherine, ate sandwiches with her during lunch, and bonded with her over an enthusiasm for gust alleviation and wake turbulence. Virtually every history of the space program would include their names—John Mayer, Carl Huss, Ted Skopinski, W. H. Phillips, Chris Kraft, and others.

Katherine Goble had stood behind the engineers' numbers for the past three years, and as humans bounded beyond the sky she would continue to do so. Like many other Americans, Katherine bridled at the reality of the Russians' metal moon orbiting overhead. *We can't let that pass without doing something about it*, she thought. But beyond sating the national pride that had been pricked by the Soviet advance, the prospect of being involved in something that was so untried, untested, and unexplored connected with Katherine's truest self. Getting the chance to figure out how to send humans into space was fortune beyond measure. As she worked with the engineers to build a course from the warmth and safety of their home to the cold void beyond, Katherine Goble's talents would truly take flight.

Dorothy Vaughan watched the furor from a second-floor office in the Unitary Plan Wind Tunnel, Building 1251. The Unitary Plan Tunnel had come online in 1955, funded by legislation to build state-of-the-art wind tunnels at each of the NACA's three main laboratories. The team that occupied most of the new building managed its own section of computers, like all the laboratory's divisions now.

Physically, Dorothy and the West Computing office had never been closer to the high-speed future. As the laboratory embraced the onset of the space age, the Unitary Plan Wind Tunnel would remain one of the busiest hubs of the center, testing "nearly every supersonic airplane, missile, and spacecraft" that saw the light of day over the next two decades. But in terms of the center's computing operations, Dorothy's pool now existed on the periphery. By 1956, more black women were now working in other areas of the laboratory than in West Computing itself. After more than a decade in their two-room spread in the Aircraft Loads Laboratory, Dorothy and the remaining women had been downsized to the new office in 1251. Miriam Mann, Ophelia Taylor, Chubby Peddrew, and many others from West Computing's class of 1943 had, like Katherine Goble and Mary Jackson, been offered permanent positions with engineering groups. Dorothy Vaughan was more likely to run into her former colleagues in the Langley cafeteria or the parking lot than to see them during the workday.

Dorothy had glimpsed the shadows of her own future when Langley disbanded the East Computing pool in 1947. Each new facility the laboratory built fueled the demand for specialization among its professionals. As the answers to the fundamental problems of flight became clearer, the next level of questioning required finer, more acute knowledge, making the idea of a central computing pool—generalists with mechanical calculating machines, capable of handling any type of overflow work—redundant. If anything, the NACA's response to Sputnik would only intensify the process of change, as the herculean task of safely navigating the heavens was divided into myriad smaller tasks, tests, parts, and people. Expertise in a subfield was the key to a successful career as an engineer, and expertise was becoming a necessity for the mathematicians and computers as well. Without it, women remaining in the segregated pool were left in a state of technical limbo.

Getting hired by the laboratory as a professional mathematician had been an important and groundbreaking stride for the black women—for all of Langley's women, of course. Their employment represented an expansion of the very idea of who had the right to enlist in the country's scientific workforce. From the beginning of the computing pools, the women easily hurdled the engineers' expectations, raising the bar as they did it. As the days of World War II receded into memory, so did the notion that riveters and gas station attendants and munitions experts and, yes, even mathematicians would, or even should, be female. And yet, away from public view, one of the largest concentrations of professional female mathematicians in the United States stayed on the job, their identities wedded to their professions.

The defense machine's hunger virtually assured them of a job through retirement. Advancement, however, would require a different plan of attack. It was a concept easily grasped, empirically proven, but far from simple to execute: if a woman wanted to get promoted, she had to leave the computing pool and attach herself to the elbow of an engineer, figure out how to sit at the controls of a wind tunnel, fight for the credit on a research report. To move up, she had to get as close as she could to the room where the ideas were being created.

With East Area Computing gone, West Area Computing was boxed in on two fronts. Not only was the group all black, it was also the only stand-alone all-female professional section left at the laboratory, and by the late 1950s, that had become an anachronism. The black men, like Thomas Byrdsong and Jim Williams and Larry Brown, certainly had to spar with racial prejudice, but they started their Langley careers with all the privileges of being a male engineer. And although the lacunae of computing pools attached to PARD and Flight Research and the plethora of tunnels were also staffed and supervised by women, those women, including the newly integrated black computers, reported directly to researchers, and they were closely tethered to the work and the status of the male engineers whose spaces they shared. Like Virginia Tucker before her, Dorothy Vaughan now presided over an appendix, still attached to the research operation but whose function had attenuated over time.

Dismantling East Computing had been a simple matter of operations, of supply and demand and expedience. When that pool's numbers grew too small to warrant maintaining a section, the laboratory simply distributed the stragglers into other sections and passed their outstanding assignments along to West Computing. But as long as "West Computer" was still the unspoken code for "Colored Computer," the decision to draw the curtain on Dorothy's group would require more nuanced consideration.

The progress that the black women had made in the last fourteen years was unmistakable. Demand for their mathematical abilities had opened Langley's front door to them, and the quality of their work had kept them at their desks. Through the familiarity that came with regular contact, they had been able to establish themselves not as "the colored girls" but simply "the girls," the ones engineers relied upon to swiftly and accurately translate the raw babble of the laboratory's fierce machines into a language that could be analyzed and turned into a vehicle that cut through the sky with grace and power.

True social contact across the races was well nigh impossible, yet within the confines of their offices, relationships cultivated over intense days and long years blossomed into respect, fondness, and even friendship. The colleagues exchanged Christmas cards with one another, asked after spouses and children. An engineer's wife gave Miriam Mann's daughter a shiny new penny to put in her shoe on her wedding day. The employees came together for extracurricular activities based at the laboratory: in 1954, Henry Reid appointed Chubby Peddrew to serve as one of the directors of Langley's inaugural United Fund Drive. The Activities Building was the site of club meetings and branch get-togethers, an end run around the embarrassment and difficulty of finding a venue in the town that would accommodate a racially mixed group. The Negro employees began attending centerwide events such as the annual Christmas party; one season, Eunice Smith volunteered as a Santa's helper. Every year, Dorothy Vaughan's children counted the days until the laboratory's giant picnic, where they could romp and play with the other kids and eat their fill of grilled hot dogs and hamburgers.

The social and organizational changes occurring at Langley were buoyed by the civil rights forces gathering momentum in the country. A. Philip Randolph, implacable in his advocacy of voting rights and economic equality, was actively working with younger organizers, principally the minister of a Montgomery, Alabama, church named Martin Luther King Jr. King and a fellow pastor named Ralph Abernathy had helped organize a boycott of the city buses after a fifteen-year-old student named Claudette Colvin and Rosa Parks, a forty-two-year-old seamstress, were both hauled off to jail for refusing to yield their seats in the "white" section of the bus. As with the legal case of Irene Morgan, the woman arrested in Virginia's Gloucester County in 1946 for the same infraction, the battle over integration on Montgomery buses eventually won a hearing in front of the Supreme Court. Once again America's highest court ruled segregation illegal. The controversy over the bus boycott vaulted the young Dr. King into the national headlines as the leader of the civil rights movement.

Langley Air Force Base and Fort Monroe moved forward to integrate the housing and the schools on their bases; as federal outposts, they were bound to comply with federal law. The state of Virginia, on the other hand, hoisted the Jim Crow flag even higher. In the years following the *Brown v. Board of Education* ruling, Senator Harry Byrd's antipathy toward the law had swelled into a countering movement—Massive Resistance—and he marshaled every resource at his political organization's disposal to build a firebreak against integration. Byrd Machine politician J. Lindsay Almond assumed the governorship and the party line in January 1958. "Integration anywhere means destruction everywhere," Almond inveighed in his inaugural address, his words a dark mirror of Lyndon Johnson's anxious commentary on Sputnik. Claiming to be the front line of defense for the entire South and its "way of life," the southern Democrats who ruled the state passed a package of laws that gave the legislature the right to close any public school that tried to integrate. "How can Senator Byrd and [Virginia] Congressman Hardy be so distressed one minute about our lagging behind the Russians in our missile program and the next minute advocate closing the schools in Virginia?" demanded one *Norfolk Journal and Guide* columnist.

Supporters of integration and segregation faced off with growing intensity: in 1956, the NAACP filed lawsuits in Newport News, Norfolk, Charlottesville, and Arlington, with the aim of forcing each of those Virginia school districts to integrate. The Byrd cronies retaliated by diverting taxpayer money to fund whites-only "segregation academies," private schools founded to circumvent integrated public schools. The no-go situation in the Virginia schools was evidence of just how difficult it was going to be to pull out the roots of the caste system that had defined and circumscribed virtually every interaction between whites and those considered nonwhite since the English first set foot on the Virginia coast. "While integration waits to be born, the 'separate but equal' education of the Negroes marks time," wrote journalist James Rorty in *Commentary Magazine*.

That so many West Computers managed to find opportunity as they rotated into new positions at the lab certainly relieved some of the pressure for Langley management to take a more active hand in the matter of integration. Langley might easily have continued its organic approach to desegregation, ending West Area Computing only after the last of the women had found a new home with an engineering section, like grade school kids waiting to be picked for a kickball team. Driven by the pragmatic sensibility of the engineers, management had naturally tacked toward a policy of benign neglect with respect to the bathroom signs and lunchrooms, neither enforcing compliance with the rules nor eliminating them altogether. It might have taken years longer before the unseen hand that had been vanquished by Miriam Mann in the lunchroom in the early 1940s would take the next step and pry the aluminum COLORED GIRLS signs off Langley's bathroom doors. But by leapfrogging the United States into space, the Russians had turned even local racial policy into fodder for the international conflict. In forcing the United States to compete for the allegiance of yellow and brown and black countries throwing off the shackles of colonialism, the Soviets influenced something much closer to Earth, and ultimately more difficult than putting a satellite, or even a human, into space: weakening Jim Crow's grip on America.

"Eighty percent of the world's population is colored," the NACA's chief legal counsel Paul Dembling had written in a 1956 file memo. "In trying to provide leadership in world events, it is necessary for this country to indicate to the world that we practice equality for all within this country. Those countries where colored persons constitute a majority should not be able to point to a double standard existing within the United States." It would take a lot more than a shiny Soviet ball and the threat of international disdain to completely break the Byrd organization's commitment to racial segregation. As far as the segregationists were concerned, racial integration and Communism were one and the same and posed the same kind of threat to traditional American values. Yet those charged with mounting the American offense in space saw strength in countering the Russian value of secrecy with its opposites— transparency, democracy, equality—and not a simulacrum.

Though many competitors within the US government were vying to lead the space effort—among them the US Air Force, the US Naval Research Observatory in Washington, DC, and Wernher von Braun and the Germans who ran the Army Ballistic Missile Agency in Huntsville, Alabama—it was the NACA that was chosen as the repository for all of America's disparate space operations. The NACA—civilian and innocuous, abundant in engineering talent—was the perfect container. In October 1958, with Mother Langley as the nucleus, the US government fused all the competing operations, along with the Jet Propulsion Laboratory, into the NACA. The expanded mission called for a new name: the National Aeronautics and Space Administration, or NASA.

The NACA was quiet, obscure, and largely overlooked. NASA would be high-profile, high-stakes, and scrutinized by the world. The work done by the NACA nuts was hidden behind the more public operations of the military services and commercial aircraft manufacturers. NASA was chartered "to provide for the widest practicable and appropriate dissemination of information concerning its activities," with all failures and tragedies of the endeavor laid bare to the citizenry and broadcast through the influential young medium of television. With the world watching, the new organization carrying the American banner into space would have to be "clean, technically perfect, and meritocratic, the bearer of a myth."

The transition from the NACA to NASA didn't change Langley's facilities significantly, nor did it require drastic changes in the laboratory's staff. But the shift in attitude and in public responsibility at the laboratory were as distinct in character as the golden age of aeronautics of the 1950s would be from the space-age 1960s. The quirky place where upstart engineers competed to "bootleg" their own projects with the knowing wink of their supervisors, where a central laboratory had grown organically into a culturally cohesive organization of five thousand had, from October 1957 to October 1958, become a high-profile bureaucracy with ten research centers and ten thousand employees.

As the Space Act of 1958 made its way through Congress, trailing behind it the sheaves of legal documents and memoranda required to bring NASA to life, one memo quietly circulated at what was soon to be renamed the Langley Research Center, authored by Langley's assistant director, Floyd Thompson, dated May 5, 1958, officially ending segregation at Langley.

"Effective this date, the West Area Computers Unit is dissolved."

As the clock ticked down on the NACA, only nine West Computers remained in the pool: Dorothy Vaughan, Marjorie Peddrew, Isabelle Mann, Lorraine Satchell, Arminta Cooke, Hester Lovely, Daisy Alston, Christine Richie, Pearl Bassette, and Eunice Smith. With one terse line of text, NASA crossed a frontier that had not been breached by its predecessor. The memo heralded the end of an era, the swan song of the Band of Sisters. The story of West Area Computing—how Dorothy

Vaughan and her colleagues found their way to Langley, the tragedy and hope of World War II, the tyranny of the signs in the Langley cafeteria and on the bathroom doors, the women's contributions to one of the most transformative technologies in the history of humankind—would get passed along as family lore, but leave barely a fingerprint in the histories of the black men and women who fought for progress in their communities, of the women who pushed for equality for their gender in all aspects of American life, or of the engineers and mathematicians who taught humans to fly. For the rest of their lives, the former West Computers reminisced with one another and with the East Computers and the engineers they worked with. They told tales at the retirement parties that crowded their calendars in the 1960s and 1970s and 1980s, but with the modesty characteristic of women of their generation, they were reluctant to describe their achievements as anything more than "just doing their jobs."

The end of the West Area Computing section was a bittersweet moment for Dorothy Vaughan. It had taken her eight years to reach the seat at the front of the office. For seven years after that she ruled the most unlikely of realms: a room full of black female mathematicians, doing research at the world's most prestigious aeronautical laboratory. Her stewardship of the section had supported the careers of women like Katherine Goble, who would ultimately receive her country's highest civilian honor for her contributions to the space program. The standards upheld by the women of West Computing set a floor for the possibilities of a new generation of girls with a passion for math and hopes for a career beyond teaching. Just as the original NACA-ites would forever hold on to their identities as members of that venerable organization, the black women would always feel an allegiance to West Area Computing, and to the woman who led it to its final day, Dorothy Vaughan.

Dorothy was forty-eight years old in October 1958, with more than a decade of work still stretching out before her. Her older children, so tiny when she had first come to Hampton Roads, were now entering college. The younger boys were adolescents following fast in the path of their older siblings. Her work at Langley had enabled her to make good on her promise to her children and their futures. With their educations on track and a house of her own in her name—the Vaughans also left Newsome Park, in 1962—there was nothing to stop Dorothy from making the final years of her career about her own ambitions.

"She was the smartest of *all* the girls," Katherine Goble would say of her colleague, years into her own retirement. "Dot Vaughan had brains coming out of her ears" (and Katherine Goble knew from brains). Dorothy was proud of the way she had navigated through the days of racial segregation, proud of whatever small share she might claim in contributing to the demise of that backward practice. She had watched the women of West Computing, along with the others at the laboratory, take flight within the NACA's research operations; together, they proved that given opportunity and support, a female mind was the analytical equal of its male counterpart. But despite knowing for many years that this day would eventually come, and having done everything within her power to bring it about, the victory she savored as the memo circulated was tempered with disappointment. Progress for the group meant a step back for its leader; Dorothy's career as a manager came to an end on the last day of the West Area Computing office.

Dorothy had never been one to linger over the past; the decade waiting in the wings promised to be one of the most interesting ever witnessed at the laboratory. For better or worse, Langley's fresh start was giving Dorothy Vaughan a fresh start as well. She would now begin life at the new agency as she had started her career at the NACA: as just one of the girls.

CHAPTER SEVENTEEN

Outer Space

This is not science fiction," wrote President Eisenhower, in the preface to a fifteen-page document entitled *Introduction to Outer Space*. Prepared by the President's Advisory Committee on Science in March 1958 as a primer on spaceflight, the brochure laid out the scientific principles of travel beyond the Earth's atmosphere in terms a layperson could understand. "As everyone knows, it is more difficult to accelerate an automobile than a baby carriage," read one passage. It also made the case for why a space program—and its enormous price tag—was in the interest of every American, offering four arguments for the public's consideration. National defense and global prestige, of course, were the two concerns that had moved the reverie of space travel from the purview of novelists and eccentrics to the country's number-one priority. The only thing that rivaled Americans' fear of the Soviet Union's incipient prowess in the heavens was their wounded national pride.

Thirdly, space exploration would bring an unprecedented opportunity to expand the body of human knowledge about the universe, prompted the pamphlet. Sputnik launched smack in the middle of the International Geophysical Year, and experts around the world fantasized over the cornucopia of data that might be harvested by a satellite or solar system–faring probe, a mechanical, electrical proxy for their own inquisitive eyes.

Katherine Goble certainly acknowledged the value of those three rationales, but for her, it was the one listed on the first page of the brochure that resonated most: humans pined to go into space because of their longing to know what lay beyond the confines of their own small world; they desired to leave Earth out of a compelling urge to go where no human had gone before. Katherine had always been driven by curiosity, and as the activity in and around Building 1244 crescendoed, it consumed her. Eisenhower's brochure put forth a vague, practically useless timetable for when the United States might be expected to achieve a variety of objectives in space: "Early," "Later," "Still Later," "And Much Later Still." The real schedule—and no one knew this better than the people in Building 1244—was As Soon As Humanly Possible. *When* America should venture beyond the confines of Earth was just as obvious as *why*. But *how*? That was what Katherine Goble ached to know.

She was far from alone. The plan for planting the American flag in the heavens, and the decision regarding who would lead the charge, was the table topic at Wright-Patterson Air Force Base in Ohio, at Wernher von Braun's Army Ballistic Missile Agency in Alabama, and at the Naval Observatory in Washington, DC. Officials gathered around conference tables at the NACA headquarters and at each of the NACA laboratories, concerned with plotting the quickest possible path into space. Nowhere vibrated with more anticipation than Langley. Katherine Johnson's deskmates—John Mayer, Ted Skopinski, Alton Mayo, Harold "Al" Hamer, Carl Huss—moved from one meeting to another, conferring with each other, with their bosses, with representatives of aircraft manufacturers and military services, turning to every possible source in order to aggregate intelligence for the still inchoate endeavor.

The only real reference that the Langley brain busters could lay their hands on was *Introduction to Celestial Mechanics*, a 1914 textbook by Forest Ray Moulton. So the engineers, who knew more about flying vehicles than any others, began scaling the next learning curve. Katherine's branch chief, Henry Pearson, organized a "self-education" lecture series that began in February 1958 and lasted through May, drafting individual engineers in Flight Research and PARD to present on one of seventeen topics related to space technology. Even in the early, confusing months after Sputnik, the top engineers in those divisions, with decades of experience in flight-test research (and many with a not-so-secret love of science fiction) sensed that they were on the cusp of a once-in-a-lifetime opportunity. They threw themselves into the class. John Mayer tackled orbital mechanics, Al Hamer lectured on rocket propulsion, and Alton Mayo handled reentry, the problems faced by an object returning to Earth. Carl Huss taught the physics of the solar system. Ted Skopinski was the trajectories guy, elaborating on the math describing the path taken by a space vehicle as it left Earth's surface and settled into orbit around it.

Katherine Goble had fallen in love with her job at Langley virtually the moment she walked through the door of West Computing. The four years she had spent doing monotonous calculations on gust alleviation had only intensified the desire to drain every drop of knowledge she could from the engineers she worked with. With the transmutation of her division's priorities from aeronautics to space, however, her work was taking a particularly toothsome turn. Massaging the Monroe calculator and filling out the data sheets, which grew longer and wider as the work became more intricate, would still be part of her daily duties. But the engineers in the group now assigned her the job of preparing the charts and equations for the well-received space technology lectures. It was like a bell sounding, taking her back to the course on the analytic geometry of space that Dr. Claytor had created for her. Claytor's demanding, rapid-fire instruction had laid the foundation both for the content of the work at hand and for its intensity. That preparation was critical as she put the abstract three-dimensional Cartesian plane to use in the service of the space technology lectures, which were eventually compiled in written form. It was the textbook of space the *place*, being written in real time.

Katherine listened carefully to everything the engineers said, strained for snippets of conversation, and devoured *Aviation Week* like a kid reading the funny papers. The real action, she knew, was taking place there in the lectures and editorial meetings, those closed-door sessions where engineers subjected preliminary research reports to the same relentless scrutiny and stress testing that they applied to the aircraft they engineered. Her interest in the proceedings of the meetings increased in direct proportion to her proximity to them. By the measure of the rest of the country, she was an insider's insider. She enjoyed a front-row seat at a spectacle that the rest of the citizenry learned about in the daily newspaper and on the nightly news. But however close she sat to the room where the meetings took place, she was still an outsider if she couldn't get in the door.

Building an airplane was nothing compared to shepherding research through Langley's grueling review process. "Present your case, build it, sell it so they believe it"—that was the Langley way. The author of an NACA document—a technical report was the most comprehensive and exacting, a technical memorandum slightly less formal—faced a firing squad of four or five people, chosen for their expertise in the topic. After a presentation of findings, the committee, which had read and analyzed the report in advance, let loose a barrage of questions and comments. The committee was brusque, thorough, and relentless in rooting out inaccuracies, inconsistencies, incomprehensible statements, and illogical conclusions obscured by technical gibberish. And that was before subjecting the report to the style, clarity, grammar, and presentation standards that were Pearl Young's legacy, before the addition of the charts and fancy graphics that reduced the data sheet to a coherent, visually persuasive point. A final report might be months, even years, in the making.

Katherine sat down with the engineers to review the requirements for the space technology lectures and the research reports that were starting to come out of the process. She listened closely to their instructions and, as was her habit, she asked questions. Not just questions designed to clarify the marching orders she had been given, but the kind of queries she had fired at her parents and teachers as a child, meant to broaden and deepen her understanding of how things work so she could create a more refined model of the world. Why did the trajectory equation need to account for the oblateness of Earth? Why was it necessary to calculate an error ellipsoid to accurately predict the satellite's return to the planet's surface?

She had asked plenty of questions when the scope of her work had extended only from the nose cone of a tiny Cessna 405 to its tail fin. Now there was so much more to ask, so much more to understand, and because it was all new, she felt like she was right there on the learning curve with the engineers. As the work intensified, something that had been hibernating in her mind awakened, and once roused it would not go away. She considered the issue and checked its logic, just as she did with her analytical work. At first she asked it only of herself, but eventually she came to the engineers with the question.

"Why can't I go to the editorial meetings?" she asked the engineers. A postgame recap of the analysis wasn't nearly as thrilling as being there for the main event. How could she not want to be a part of the discussion? They were her numbers, after all.

"Girls don't go to the meetings," Katherine's male colleagues told her.

"Is there a law against it?" Katherine retorted. There wasn't, in fact. There were laws telling her where she might answer nature's call—a law she ignored at Langley—and which fountain to drink from. There were laws restricting her ability to apply for a credit card in her own name, because she was a woman. But no law applied to the editorial meeting. It wasn't personal: it was just the way things had always been done, they told her.

Restricting the computers from joining the editorial meetings wasn't a rule: it was a rule of thumb. It was rooted in practice and widely implemented, but it did not apply without exception to every situation. Langley gave each division chief, and every branch head and section head below them on the ladder, a certain amount of leeway in the management of their groups. Whether or not a woman was promoted, if she was given a raise, if she had access to the smoky sessions where the future was being conceived and built, had much to do with the prejudices and predilections of the men she worked for.

In 1959, six of Langley's female employees—Lucille Coltrane, Jean Clark Keating, Katherine Cullie Speegle, Ruth Whitman, Emily

Stephens Mueller, and Dorothy Lee—assembled around a table in a Langley office to sit for a group photo, their elegant, well-made suits amplifying the confidence in their gaze. "Women Scientists," the photographer labeled the picture, though the particulars of the occasion would be lost to the passage of time. They had rated inclusion in the photograph because of some combination of rank, research contributions, and general esteem in the eyes of their bosses. Five out of the six women in the photo worked in PARD.

One of the women in the photo, Dorothy Lee, had accepted a position as a computer in PARD in 1948, fresh out of Randolph-Macon Women's College in Virginia, just after East Computing was disbanded. When branch chief Maxime Faget's secretary took off for a two-week honeymoon, Dorothy was asked to sub for her. She answered the phones and distributed the mail in addition to her regular duties, which at the time involved solving a triple integral for an engineer in the division. At the end of the two weeks, she had so impressed Faget with her math (not her secretarial skills, as she didn't know how to type) that he invited her to become a permanent member of his branch, apprenticing her to men who showed her the ropes of aerodynamic heating. By 1959 she had authored one report, coauthored seven more, including one with Max Faget, and, like Mary Jackson, been promoted to engineer.

Early in her career at Langley, Dorothy Lee was interviewed for the *Daily Press*, in all probability by Virginia Biggins, the female reporter assigned to the Langley beat. "Do you believe," she was asked, "that women working with men have to think like a man, work like a dog, and act like a lady?" "Yes, I do," Lee said, who was then mildly mortified to read her words in the Sunday paper.

It was the "acting like a lady" term of the equation that was so vexing. A little bit of coyness, like wormwood, could be pleasantly intoxicating, smoothing interactions with the men. Too much politesse, however, might poison a woman's prospects for advancement. Women were "supposed" to wait for the assignments from their supervisors, and weren't expected to take the lead by asking questions or pushing for plum assignments. Men were engineers and women were computers; men did the analytical thinking and women did the calculations. Men gave the orders and women took notes. Unless an engineer was given a compelling reason to evaluate a woman as a peer, she remained in his blind spot, her usefulness measured against the limited task at hand, any additional talents undiscovered.

Some women did indeed spend their days in rote service to the day's task, plotting data with blithe indifference, routing torrents of numbers as nonchalantly as the calculating machines they cradled. But the average level of interest in the work among female employees was no lower than it was for their male counterparts, the "inveterate wind tunnel jockeys" and the mediocre "can't-hack-it engineers" who managed to carve out a comfortable place for themselves in the bureaucracy despite modest talents or ambition. For the women who had found their true calling at the NACA, like Dorothy Lee, like Katherine Johnson, they woke up dreaming of angles of attack and two-body orbit equations and ablation processes no less than did Chris Kraft and Max Faget and Ted Skopinski. They matched their male colleagues in curiosity, passion, and the ability to withstand pressure. Their path to advancement might look less like a straight line and more like some of the pressure distributions and orbits they plotted, but they were determined to take a seat at the table. First, however, they had to get over the high hurdle of low expectations.

Whatever personal insecurities Katherine Goble might have had about being a woman working with men, or about being one of the few blacks in a white workplace, she managed to cast them aside when she came to work in the morning. The racism stuff, the woman stuff: she managed to tuck all that way in a place far from her core, where it would not damage her steely confidence. As far as Katherine was concerned—as far as she had *decided*—once they got to the office, "they were all the same." She was going to assume that the smart fellas who sat across the desk, with whom she shared a telephone line and the occasional lunchtime game of bridge, felt the same. She only needed to break through their blind spots and make her case.

"Why can't I go to the editorial meetings?" Katherine Goble asked again, undeterred by the initial demurral. She always kept up the questioning until she received a satisfactory answer. Her requests were gentle but persistent, like the trickle of water that eventually forces its way through rock. The greatest adventure in the history of humankind was happening two desks away, and it would be a betrayal of her own self-confidence and of the judgment of everyone who had helped her to reach this point to not go the final distance. She asked early, she asked often, and she asked penetrating questions about the work. She asked with the highest respect for the natures of the brainy fellas she worked with, and she asked knowing that she was the right person for a task that needed the finest minds.

As much as anything, she asked with confidence in the ultimate decision.

"Let her go," they finally said, exasperated. The engineers just got tired of saying no. Who were they, they must have figured, to stand in the way of someone so committed to making a contribution, so convinced of the quality of her contribution that she was willing to stand up to the men whose success—or failure—might tip the balance in the outcome of the Cold War?

In 1958, Katherine Goble finally made it into the editorial meetings of the Guidance and Control Branch of Langley's Flight Research Division, soon to be renamed the Aerospace Mechanics Division of the nearly-ready-for-prime-time National Aeronautics and Space Administration. Now, she was going to come along with the program.

CHAPTER EIGHTEEN

With All Deliberate Speed

1958 was a year no Langley employee would ever forget. Leaving work on September 30, they said good-bye to the National Advisory Committee for Aeronautics, the esoteric operation that for forty-three years had quietly supervised and directed the airpower revolution, good-bye to the Langley Aeronautical Laboratory of yore. On the morning of October 1, the former NACA-ites walked into the Langley Research Center, epicenter of the National Aeronautics and Space Administration, a new American agency whose birth had been induced by a hurtling Soviet sphere. The buildings hadn't changed, nor had the people, or, for many of them, the work they were charged with. But from sundown one day to sunup the next, they had gone, if only in the public imagination, from erudite and obscure to obvious and spectacular, from the crackpots of the airplane epoch of the 1940s to the guardians of the space-age 1960s.

At the end of the 1950s, when the American space program looked as uncoordinated and spindly as a foal, predicting that the United States would best the Soviets might have seemed like a fool's bet. NASA had other plans, creating a brain trust at Mother Langley called the Space Task Group, a nimble, semi-autonomous working group that drew largely from the Flight Research Division and PARD and was led by engineer Robert Gilruth. The Space Task Group set up shop on Langley's East Side in some of the laboratory's oldest buildings. Those space pilgrims, an initial group of forty-five people, gave the country's first manned space program an operating plan and a name: Project Mercury. The venture had three goals: to orbit a manned spacecraft around the Earth, to investigate man's ability to function in space, and to recover both men and spacecraft safely.

Virginians puffed out their chests with pride now that the good old brain busters were leading the charge against the Reds. An October 1959 open house at Langley held on the occasion of NASA's first anniversary attracted twenty thousand ardent locals eager for an up-close look at the work of the unusual neighbors they had underestimated and overlooked for decades. No longer just a "a dull bunch of gray buildings with gray people who worked with slide rules and wrote long equations on blackboards," NASA, the public now believed, was all that stood between them and a Red sky. However, Virginia's legacy as the birthplace of humanity's first step into the heavens would have to compete with the notoriety it was gaining as the country's most intransigent foe of integrated schools.

"So far as the future histories of this state can be anticipated now, the year 1958 will be best known as the year Virginia closed the public schools," lamented Lenoir Chambers, editor in chief of Virginia Beach's *Virginian-Pilot* and a southern liberal in the mold of Mark Etheridge of the *Louisville Post-Courier*. Undeterred and unchastened by the 1957 showdown in Little Rock, the Byrd Machine's Massive Resistance movement made good on its threat. In the fall of 1958, Virginia's governor Lindsay Almond chained the doors of the schools in localities that attempted to comply with the Supreme Court's *Brown* decision. Thirteen thousand students in the three cities that had moved forward with integration—Front Royal, Charlottesville, and Norfolk—found themselves sitting at home in the fall of 1958. "I would rather have my children live in ignorance than have them go to school with Negroes," one white parent told a reporter. A total of ten thousand of the shut-out students lived in Norfolk: 5,500 of those from military families stationed at the naval base, white students as well as black paying the price for the state's racial crusade.

Across the water from Norfolk, on the peninsula that Langley called home, public schools remained open but segregated. Even as the barriers in their parents' workplace continued to erode, the children of Langley's black employees returned to their fall routines at Carver, Huntington, and Phenix, while their white colleagues' children went back to Newport News High and Hampton High. In their new home in Mimosa Crescent, the Goble daughters were now zoned to attend Hampton High School. The school board, however, paid "school fees" to the families as an incentive for them to keep their children in the black district, similar to the out-of-state "scholarships" the state offered to black graduate students to keep them from integrating Virginia colleges.

The forces in favor of equality redoubled their efforts, determined to surmount the resistance to integration like a jet engine propelling an airplane through drag. But, like Christine Mann and everyone else whose hopes—and fears—had escalated on the day the *Brown* case was decided, blacks in Virginia were acutely aware of the long lag between legal and political triumphs and social change. As fantastical as America's space ambitions might have seemed, sending a man into space was starting to feel like a straightforward task compared to putting black and white students together in the same Virginia classrooms.

Rather than trying to make plans based on machinations beyond their reach, parents like Dorothy Vaughan, Mary Jackson, and Katherine Goble worked hard to influence what they did control: pushing their children to excel in their segregated schools and getting them into college. Katherine Goble's eighteen-year-old daughter, Joylette, a talented violinist and a graceful beauty, graduated salutatorian of Carver High School's class of 1958 and headed across town to attend Hampton Institute. Connie and Kathy, honor students and musicians in Carver High's sophomore class, nipped at their elder sister's heels. The girls and their mother made regular appearances in the social column of the *Norfolk Journal and Guide*, the model of an upwardly mobile and professional black family.

In public, Katherine Goble was unfailingly gracious, optimistic, and unflappable, and she insisted that her girls acquit themselves in the same fashion. Her grief and loneliness, the burden of being both mother and father, she relegated to the privacy of their house on Mimosa Crescent. Jimmy Goble had been the love of Katherine's youth, a nurturing father, and the partner she expected to grow old with. The two of them had been a compatible, attractive, and charming couple, making the rounds of the black community's fall galas, debutante balls, picnics, and fundraisers. As a single woman, still youthful at forty years old, she found herself drifting toward the social sidelines.

Eunice Smith was Katherine's steadfast companion and confidante. The two of them spent more time together than many married couples, commuting back and forth to work each day, serving together as officers of the Newport News chapter of their sorority, AKA, taking time off from work to root for their teams in the yearly Central Intercollegiate Athletic Association (CIAA) basketball tournament for black colleges. They never missed Sunday service at Carver Memorial Presbyterian Church, and one night a week when they left Langley they headed over to Carver for choir practice.

One evening in 1958, a handsome thirty-three-year-old army captain with a ready smile and a rich bass voice ambled into practice. James A. Johnson, born in rural Suffolk, Virginia, had moved with his family to Hampton as an adolescent. He attended Phenix High School, and in fact Mary Jackson had been one of his student teachers. Jim Johnson had planned to attend Hampton Institute but was drafted right after graduating from high school. Rather than being assigned to the US Naval Training School there on the campus, he was sent to the US Navy Boot Camp in Great Lakes, Illinois. He trained in aviation metalsmithing, specializing in the repair of propellers. After his war service, Johnson finished his degree and landed a clerk job at the Commerce Department in Washington, DC, but he also signed up for the US Navy Reserve so he could spend his weekends at Patuxent River Naval Base in Maryland, repairing planes used for test flight. With the onset of the Korean War, he enlisted in the army, serving as an artillery sergeant, calibrating guns being fired on enemy infantry. In 1956, he returned to

Hampton, taking a job at the post office as a mail carrier, maintaining his trim military shape through miles of walking each day. Never one to stray too far from the armed services, he also signed up as a member of the US Army Reserve.

"Ladies, he's single," the pastor had announced in church that Sunday after introducing Jim as a new member of the congregation. It hadn't been Katherine's expectation or intention to find a new love, but almost immediately after meeting in the choir loft, she and Jim began courting, tentatively turning up together at dances and dinner parties and arriving together at church as a family, with Kathy and Connie in tow.

Jim's devotion to the military service made it easy for him to understand Katherine's strong commitment to her work at Langley. He knew the satisfaction that came from fulfilling employment and loved the sense of mission and camaraderie that the military gave him. As a black man, he relished the opportunity to step forward from the cook and steward and laborer jobs that had traditionally been reserved for blacks and gain expertise in an area where he felt he could make a frontline contribution.

He was also sensitive to the secretive nature of Katherine's work and the longer hours her job now demanded of her. Since the end of World War II, the NACA had been an eight-to-four-thirty kind of place. Now, at the outset of the space race, leaving the building at ten o'clock would be a good night. In a less urgent scenario, NASA personnel might have taken a more NACA-like approach to the problem of space by conducting a careful, measured investigation of all possible options for space travel and recommending the ones with the greatest long-term potential. There were those within NASA who believed, and would continue to believe for decades into the future, that the government's decision to put all its chips on a short-term strategy to beat the Soviets came at the cost of the opportunity to turn humans into a truly spacefaring species. With the Russians off to what looked like a commanding lead, it was the simplest, fastest, and most reliable approach that began to take shape as NASA teased out the limitations, interdependencies, contingencies, and unknowns they faced. The engineers approached Project

Mercury the way engineers tackled any problem: they broke Project Mercury down into its constituent parts.

The spacecraft itself, the can that would take a man into space, was the brainchild of Dorothy Lee's boss, Maxime Faget. Aerodynamic theory and intuition suggested that the rocket and spacecraft combination should be as streamlined as possible, to minimize aerodynamic drag. Since the Wright Brothers' 1915 Flyer, airplanes had evolved from pelican-like awkwardness to sleek machines with the silhouette of a falcon; why wouldn't a spaceship continue along that same path? But tests by Harvey Allen, an engineer at the NACA's Lewis Flight Propulsion Laboratory in Cleveland, showed that needle-shaped structures wouldn't be able to deflect the extreme heat caused when they zoomed through the friction of the atmosphere. A blunt-shaped body—something shaped more like a champagne cork—would create a shock wave as it came back toward Earth, dissipating the heat and keeping (they hoped) the man inside safe. Faget put Allen's insight to work in the design of the Mercury space capsule, six feet wide and nearly eleven feet long, weighing three thousand pounds.

The selection process for astronauts would be limited to candidates small enough to fit into the lunchbox of a spaceship: only men at or under five feet, eleven inches tall and weighing less than 180 pounds were considered. Each was required to be a qualified test pilot under forty years old with at least a bachelor's degree. In 1959, NASA held a press conference to present the "Mercury Seven" astronauts to the world. Four of the seven selected—Alan Shepard, Scott Carpenter, Wally Schirra, and John Glenn—had graduated from the US Naval Test Pilot School at Patuxent River, where Katherine's new beau, Jim Johnson, had worked as a mechanic. NASA installed the astronauts in an office at Langley next door to the Space Task Group and proceeded to put them through physical training and classroom instruction in engineering and astronautics. Employees stayed alert to catch a glimpse of the Mercury Seven, who had gone from anonymous military men to among the most recognizable faces in the world. Computers working in the Space Task Group and the astronauts, whose office was located in the same building, often ran into each other going to and coming from the bathrooms.

The rockets NASA needed to blast spacemen and spacecraft into space would come from the army's existing inventory of Redstone and Atlas missiles, overseen by Wernher von Braun at NASA's Marshall Space Center in Huntsville, Alabama. The propulsion experts at NASA's laboratory in Cleveland took the lead on the craft's electrical system and the retrofire rockets built into the craft itself.

To the engineers on Katherine's desk fell the responsibility of the trajectories, tracing out in painstaking detail the exact path that the spacecraft would travel across Earth's surface from the second it lifted off the launchpad until the moment it splashed down in the Atlantic. As the head of the Space Task Group, Robert Gilruth had been given his pick of NASA employees to fill the ranks of Project Mercury's nerve center. Katherine's office mate, John Mayer, had jumped ship for the new endeavor a week after it was created, in November 1958. The workload generated by Project Mercury was so onerous that even after Mayer transferred from 1244 to the offices on the East Side, he "bootlegged" the overflow work to his old buddies Carl Huss and Ted Skopinski, getting them to help out with whatever time they could squeeze in around what they owed to Henry Pearson. He got them to do "computing runs" for him—which meant getting Katherine to do computing runs for them. The group took on the additional tasks with zeal, because space looked like "a hell of a lot of fun." They turned their desks into a trigonometric war room, poring over equations, scrawling ideas on blackboards, evaluating their work, erasing it, starting over.

There was virtually no aspect of twentieth-century defense technology that had not been touched by the hands and minds of female mathematicians. Like Katherine and her colleagues at Langley, women at the Aberdeen Proving Ground in Maryland spent thousands upon thousands of woman-hours computing ballistics trajectory tables, which soldiers used to accurately calibrate and fire their weapons, as Jim Johnson had in Korea. The first attempt to put a man into space, NASA decided, should be a simple ballistic flight, with the capsule fired into space by a rocket like a bullet from a gun or a tennis ball from a tennis ball machine. Capsule goes up, capsule comes down, its path defined by a big parabola, its landing place the Atlantic Ocean. The astronaut needed to return near enough to waiting navy ships to be quickly hoisted out of the water and pulled to safety. The challenge was to rig the machine's position so that the ball—the Mercury capsule—landed as closely as possible to the navy's waiting racket. Calculated incorrectly, the ball would go out of bounds, the astronaut's life endangered. The math had to be as precise and accurate as an Althea Gibson serve.

A well-executed suborbital flight would buy the United States a little breathing room; but orbital flight—the end game of Project Mercury—was infinitely more complex. Successful orbital flight required the engineers to adjust the tennis ball machine's chute to the correct angle and arm its launcher with enough force to send the ball up through the atmosphere and into an orbit around Earth on a path so precisely specified, so true, that when it came back down through the atmosphere, it was still within spitting distance of the navy's waiting racket.

"Let me do it," Katherine said to Ted Skopinski. Working with Skopinski as a computer (or "math aide," as the women had been renamed when the NACA became NASA), she had proven herself to be as reliable with numbers as a Swiss timepiece and deft with higher-level conceptual work. She was older than many of the space pilgrims, some of whom were just out of college, but she matched them at every turn for enthusiasm and work stamina. The fellas were putting everything they had on the line, and she was not going to be left out. "Tell me where you want the man to land, and I'll tell you where to send him up," she said.

Her grasp of analytical geometry was as good as that of the guys she worked with, perhaps better. And the unyielding demands of Project Mercury and the sprawling, still-forming organization that was being built to manage it stretched everyone to the limit. Soon after John Mayer put on the Space Task Group jersey, Carl Huss and Ted Skopinski followed suit, making Katherine the natural inheritor of the research report that would describe Project Mercury's orbital flight. As had been the case many other times in her life, Katherine Goble was the right person in the right place at the right moment.

Sitting in the emptier office, she plunged into the analysis, although the pesky laws of physics turned an afternoon of rote celestial tennis practice into a forces free-for-all. Earth's gravity exerted its force on the satellite and had to be accounted for in the trajectory's system of equations. Earth's oblateness—the fact that it was not perfectly spherical but slightly squat, like a mandarin orange—needed to be specified, as did the speed of the planet's rotation. Even if the capsule were to shoot off into the air directly overhead and come back down in the same straight line, it would land in a different spot, because Earth had moved.

"In the recovery of an artificial earth satellite it is necessary to bring the satellite over a preselected point above the earth from which the reentry is to be initiated," she wrote. Equation 3 described the satellite's velocity. Equation 19 fixed the longitude position of the satellite at time T. Equation A3 accounted for errors in longitude. Equation A8 adjusted for Earth's west-to-east rotation and oblation. She conferred with Ted Skopinski, consulted her textbooks, and did her own plotting. Over the months of 1959, the thirty-four-page end product took shape: twenty-two principal equations, nine error equations, two launch case studies, three reference texts (including Forest Ray Moulton's 1914 book), two tables with sample calculations, and three pages of charts.

The rapidly growing Space Task Group was taking shape as an autonomous unit marching out in front of the space parade. The new endeavor consumed as many person-hours as it could be given. Even as the Space Task Group worked to create boundaries with the research center that had given birth to it, Space Task Group employees still had responsibilities to their old managers. Katherine's and Ted Skopinski's Azimuth Angle report was the work of the Flight Research Group, the responsibility of their branch chief, Henry Pearson, and while Ted Skopinski was increasingly out of sight, spending time over at the STG offices on the East Side, the report, still unfinished, was not out of Henry Pearson's mind.

"Katherine should finish the report," Skopinski said to Pearson. "She's done most of the work anyway." Henry Pearson had the reputation of being less than supportive of the advancement of female employees, but whether it was circumstance, the triumph of hard work over bias, or an incorrectly deserved reputation, it was on his watch that Katherine put the finishing touches on her first research report on the Friday after Thanksgiving 1959. "Determination of Azimuth Angle at Burnout for Placing a Satellite over a Selected Earth Position" went through ten months of editorial meetings, analysis, recommendations, and revisions before publication in September 1960—the first report to come out of Langley's Aerospace Mechanics Division (or its predecessor, the Flight Research Division) by a female author. Stepped on, turned out, pulled apart, and subjected to every stress test the editorial committees could throw at it, Katherine's road map would help lead NASA to the day when the balance of the space race was tipped in favor of the United States.

For Katherine, the report commemorated the beginning of a new phase not just at Langley but in her personal life. Somehow, during the long, bleary-eyed days of 1959, she accepted an offer even more enticing than being invited into the editorial meetings: Jim Johnson's marriage proposal. The two married in August 1959 in a quiet ceremony at Carver Memorial. When she signed her first research report, she used a new name, the name that history would remember: Katherine G. Johnson.

CHAPTER NINETEEN

Model Behavior

Mary Jackson scrutinized every aspect of the model—its smoothness, its symmetry and alignment, its weight distribution—her trained eye and intuition sensitive to anything that might lower its aerodynamic fitness. This had been a project of nights and weekends, but she knew this investigation would provide results much more quickly than any research currently underway in the Four-Foot Supersonic Tunnel. The bar had been set the year before by an engineer in the Aerospace Mechanics Division—Katherine Johnson's group—but Mary and her young collaborator were more than up to the challenge. She was ready to spend all the time it took to help her son, Levi, build a humdinger of a car to race in the peninsula's 1960 soap box derby.

Since the beginning of the year, Mary had spent hours, hundreds of them, perhaps, collaborating with her thirteen-year-old son the way she worked with Kazimierz Czarnecki. She and Levi had gone to the local Chevrolet dealer to fill out the entry form and pick up a copy of the official rules, which read like a familiarization manual for an airplane. "The car and driver together must weigh less than 250 pounds. Only rubber wheels allowed. Length shall not exceed 80 inches. Road clearance must be at least three inches with the driver in the car. The total cost of the car shall not exceed $10.00, exclusive of wheels and axles."

They absorbed the restrictions and made sketches and measurements, trying out different designs until they settled on the best specification. Then they hunted for the materials that would bring their sketch to life. Buried in the clutter at the back of the garage might be a treasure in disguise: vegetable crates, plywood, orphaned wagon wheels, garden tools, old shoes, wire and twine—just about anything could prove useful to building the car, given enough creativity. Gluing, nailing, screwing, and fitting ensued as the big race, held annually over the July Fourth holiday weekend, approached. Mary helped her son refine the vehicle until they possessed something that could roll down the street with its pilot, as the racers were known, in the driver's seat.

The final step was to smooth, sand, and polish the body of the car to within an inch of its home-built life. All the derby's matches started at the top of a hill, with no pushing allowed. Levi and his competitors would set off from the Twenty-Fifth Street Bridge in Newport News, virtually the only thing that could pass as a hill in the flat-as-a-pancake coastal terrain. As the pilots released their brakes, they hunched themselves as far down as they could into their vehicle's cockpit, imploring the gods of gravity to pull them as quickly as possible down the nine-hundred-foot racecourse, hoping to do righteous battle against air resistance, which was as much the foe of the pee-wee racer as it had been for Chuck Yeager. No one knew that better than Levi's technical consultant, who managed to sneak in the occasional sponsored moment about the wonders of a career in the sciences in the midst of the building fun.

An enduring symbol of American boyhood (girls weren't allowed to race until the early 1970s), the All-American Soap Box Derby mixed good old American whiz-bang ingenuity with family fun. The competition had started as a Depression-era distraction, a way to create something out of nothing when nothing was what most people had. Over the years, it had taken hold at the grassroots, and in 1960 Levi was one of fifty thousand boys gearing up to compete in local races around the country. Not surprisingly, the peninsula embraced the competition with zeal. Parents who spent their days designing, building, fixing, and operating machines of transportation signed their sons up and gave free rein to their own tinkering instincts. They got to spend time with their children and let the parental mask slip just a bit, giving their offspring a glimpse of the curious child they themselves had once been. Officially, the derby was the boy's show, from building the car to crouching inside it on race day. Parents (usually fathers; Mary was one of the very rare derby moms) were supposed to stand back and offer only advice, but it was usually hard to tell who savored the engineering project more, the parent or the child.

Like craftsmen in a medieval guild, the NASA engineers hoped that one day their children would decide to take up the mantle of the profession they held so dear. Their workplace was pleasant and safe, their colleagues were smart and interesting, and over the course of the twentieth century, engineers had seen the fruits of their labor transform every aspect of modern life in ways that seemed unimaginable even as they were happening. They wouldn't get rich, but an engineer's salary was more than enough to crack into the ranks of the comfortable middle class. So they served as laboratory assistants for science projects and turned the kitchen table into an honors calculus class. They held their offspring captive until the last homework problem was solved correctly, adolescent insolence and tears be damned.

No NASA father had anything on Mary Jackson. Building a soap box derby car was an apprenticeship in engineering, and the earlier a kid got started, she knew, the more likely they were to fall under its spell. She pushed Levi (and his teachers) to allow him to take the most challenging math and science classes he could handle, and she coached him on his science projects. His eighth-grade project, "A Study of Air Flow in Scaled Dimensions," scored third place in his school's annual science fair.

"Soapbox what?" some neighbors and Bethel AME parishioners and Girl Scout troop members had asked when Mary told them about her and Levi's mechanical exploits. The first challenge many blacks faced in participating in something like the All-American Soap Box Derby was finding out about it in the first place. Starting early in the year, Chevrolet placed advertisements in *Boys' Life* magazine, the official publication of the Boy Scouts, exhorting youngsters to put in their bid for fun, fame, and adventure by getting their cars in tip-top shape before racing season rolled around in the summer. Levi, who was a member of Bethel AME's Boy Scout troop, might have read about the derby even if it hadn't been part of the watercooler conversation at his mother's office, but the message had a hard time finding its way to less well-connected ears.

Harder than getting the message, perhaps, was acting on it when you got it. Entering the derby was tantamount to believing you had a shot at victory, as much (or more) for the parents as for the racer. The electrified fence of segregation and the centuries of shocks it delivered so effectively circumscribed the lives of American blacks that even after the current was turned off, the idea of climbing over the fence inspired dread. Like the editorial meetings in 1244, like so many competitive situations large and small, national and local, black people frequently disqualified themselves even without the WHITES ONLY sign in view. There was no rule keeping a Negro boy from entering the race, but it took a lot of gumption for him to believe that he might win, and even more to accept a loss as a failure that had nothing to do with his race.

Mary, however, was determined to clamber over every fence she encountered and pull everyone she knew behind her. The deep humanitarianism that was her family inheritance had taught her to see achievement as something that functioned like a bank account, something you drew on when you were in need and made deposits to when you were blessed with a surplus.

Langley, full of talented people with varied interests, was a bonanza of recruits for her many volunteer activities. Coworkers got used to finding Mary standing quietly at their desks, enlisting them in her latest attempt to apply the engineer's values of discipline, order, and progress to the social sphere. Girls, she believed, needed particular attention; it wasn't lost on her that the derby, while open to black boys, would have rejected her daughter's application because of her gender. Mary's promotion to engineer gave her an unusual vantage point. Despite the relatively large group of women now working at the center, most female technical professionals, black and white—even someone as gifted as Katherine Johnson—were classified as mathematicians or computers, ranked below engineers and paid less, even if they were doing the same work.

Mary made common cause with the black employees working at Langley and at other places in the industry. She and Katherine Johnson and many others were core members of the National Technical Association, the professional organization for black engineers and scientists. Mary made every effort to bring students from Hampton's public schools and from Hampton Institute into the Langley facilities for tours, to get an up-close and personal look at engineers at work. She organized an on-site seminar for career counselors at Hampton Institute so that they might better steer their students into job opportunities at Langley. If she got word that Langley was hiring a new black employee, she went out of her way to make phone calls to find him or her a place to live, just as she had done when she was secretary of the King Street USO.

But Mary also cultivated allies among the white women she worked with. Emma Jean Landrum, another member of Langley's tiny engineering sisterhood, sat a couple of desks away from Mary in the Four-Foot Supersonic Pressure Tunnel office. Emma Jean was valedictorian of the University of North Carolina at Greensboro's class of 1946, working her way through school serving meals in the dining hall and grading papers for professors. Like so many of the women at Langley, Emma Jean had been recruited by Virginia Tucker, Langley's erstwhile Head Computer. In the years since, Emma Jean had produced several research reports as a part of the Unitary Plan Tunnel team; she then transferred to the Four-Foot SPT office, where she became another of Kaz Czarnecki's frequent collaborators. She, like Mary Jackson, had become an engineer in 1958.

When Mary asked Emma Jean to participate in a career panel in 1962, organized by the local chapter of the National Council of Negro Women, she readily agreed. An all-black group of junior high school girls paid close attention to Mary and Emma Jean's joint lecture, entitled "The Aspects of Engineering for Women." Afterward, Emma Jean entertained the girls with a slideshow from a trip she had recently taken to Paris and London. Their appearance together in front of the group—Mary, petite and black, and Emma, white and nearly a foot taller—made as powerful a statement on the possibilities of the engi-neering field as their actual presentation. Not only did the girls receive firsthand evidence that women could succeed in a traditionally male field; in Mary and Emma's collaboration, they saw that it was possible for a white workplace to embrace a woman who looked like them.

Serving as the leader of Girl Scout Troop No. 60, now one of the larg-est minority troops on the peninsula, was always at the top of Mary's list of volunteer activities. However, she was becoming impatient with the segregation that mandated a separate council for black scouts, and she began campaigning for one organization overseeing all the scouts. When nominations circulated to fill Virginia's two slots for the Girl Scouts' national conclave in Cody, Wyoming, Mary lobbied to send her young assistant troop leader Janice Johnson, who had developed into a capable right hand and a leader in her own right. This would be Janice's first time in an integrated setting—her first time away from her home-town, in fact—but Mary believed she would be up for the challenge and find it an invaluable experience.

Mary also knew that a native of a place so flat it was practically un-derwater would need a leg up before hiking for days in the rarefied altitude of the Wyoming mountains. So Mary enlisted the help of Helen Mulcahy, a former East Computer who had transferred to Langley's technical editing department. Mary asked Helen, an aficionado of the outdoors, to take Janice trekking with a full backpack, first on Buckroe Beach, then up into Virginia's Shenandoah mountains. It wasn't exactly the most rigorous training for an excursion at five thousand feet, and Janice didn't earn any badges for her hiking, but she held her own in the camp and returned with tales for her young charges and a head full of dreams of a life beyond her Tidewater home.

With each passing year, it seemed that the work Mary loved and the community service that gave her life meaning were becoming one and the same. She earned her engineering title through hard work, talent, and drive, but the opportunity to fight for it was made possible by the work of the people who had come before her. Dorothy Vaughan had had a positive impact on her career and on the phenomenon-in-waiting that was Katherine Johnson. Dorothy Hoover had shown that a black

woman was capable of the highest level of theoretical aeronautical research. Pearl Young, Virginia Tucker, Kitty Joyner—Mary stood on
those white women's shoulders too. Each one had cracked the hole in
the wall a little wider, allowing the next talent to come through. And
now that Mary had walked through, she was going to open the wall as
wide as possible for the people coming behind her.

On Saturday morning, July 3, an enthusiastic crowd of four thousand
people crowded along both sides of Twenty-Fifth Street in Newport
News, kicking off their Fourth of July holiday weekend. The weather was
Virginia summer at its best: clear, warm, just enough of a breeze to keep
the crowd from overheating, not too breezy to interfere with the outcome of the peninsula's tenth annual soap box derby. Contestants for
the first heat of the day wheeled their vehicles to the starting line at the
summit of the Twenty-Fifth Street Bridge. Everything receded into the
background as the pilots settled into their cars—the view of the C&O
piers and the shipyard below, the sound of the energetic crowd, the faces
of family and friends who had come to cheer them on—everything
except the feel of the vehicle confining their gangly limbs and the desire
to be the first car to cross the finish line. Officials weighed and inspected
each car and then held a lottery to determine the positions in the first
heat. At the crack of the starter's pistol, the pint-sized pilots released their
brakes, hunched down into their homemade roadsters, and willed their
cars down the hill. The race was an all-day affair, heat upon heat of
anxious and eager adolescent boys soldiering on through wobbly wheels,
broken axles, driver error, parental disappointment, and photo finishes.

Mary Jackson could see the air moving around the racer just as clearly
as if she were looking at a Schlieren photograph taken in a wind tunnel.
Levi's car was well made; the only adjustment it required between heats
was "a drop of oil on each wheel bearing." Mary and Levi Sr. and four-
year-old Carolyn held their breath as Levi Jr. got into position for the
final heat. It seemed like an eternity, but at the end, Mary and Levi Sr.
shouted in delight: their son had finished first, saving his best time for
the heat that mattered most. Wearing a black-and-white crash helmet

and the official race T-shirt, Levi Jr. sailed across the line at a relatively blazing seventeen miles per hour. His family fell upon him in a crush of hugs and celebration. To the inquiring and surprised local reporters who came to hear from the winner of the Virginia Peninsula Soap Box Derby, Levi Jackson confided the secret of his victory: the slimness of his machine, which helped to lower the wind resistance. *What do you want to be when you grow up?* the *Norfolk Journal and Guide* reporter must have asked. "I want to be an engineer like my mother," Levi said.

The spoils of the win were eye-popping: a golden trophy, a brand-new bicycle, and a spot at the national All-American Soap Box Derby in Akron, Ohio, as the official representative of the Virginia Peninsula. There Levi would face off against pilots from around the country, in front of seventy-five thousand fans, on a track where the racers could dash along at speeds exceeding thirty miles per hour. There he would be the only occupant of his aerodynamic buggy, but he'd have a community of people riding along on his shoulder. Levi Jackson was the "first colored boy in history" to win the peninsula's soap box derby. Virtually the moment he crossed the finish line, the donations started rolling in from the Bachelor-Benedicts, the Phoebus Elks, the Beau Brummell Social Club, the Hampton Women's Service League, half a dozen local black-owned businesses, and each of Hampton's three largest black churches to help defray the costs of the local hero's trip to Ohio. Another Black First for the books! If a black kid could take home the soap box derby trophy, what else might be possible?

Achievement through hard work, social progress through science, possibility through belief . . . when Levi reached out and took hold of the first-place trophy, Mary witnessed, in one proud and emotional moment, the embodiment of so much that she held dear. Of course, Mary also knew that her son was a ringer; the two of them had been building to win. Brain busters' kids were *supposed* to come out on top in a race like this, even if the brain buster was a woman, or black, or both. Being part of a Black First was a powerful symbol, she knew just as well as anyone, and she embraced her son's achievement with delight. But she also knew that the best thing about breaking a barrier was that it would never have to be broken again.

CHAPTER TWENTY

Degrees of Freedom

In February 1960, as NASA pushed forward with reliability tests on the Mercury capsule, four students from North Carolina Agricultural and Technical, a black college in Greensboro, North Carolina, sat down at the segregated lunch counter in the town's Woolworth's and refused to move until they were served. The following day, the "Greensboro Four" had become a group of twenty activists. On the third day, sixty students converged upon the Woolworth's, and by the fourth, three hundred had joined the demonstration. Participating were students from Bennett College, an all-black women's college in Greensboro, as well as white students from Guilford College and the Women's College of the University of North Carolina, the alma mater of many East Computers. Within a week, the protests, inspired by the nonviolent actions of India's Mahatma Gandhi, spread to other cities in North Carolina, and then crossed the borders into Kentucky, Tennessee, and Virginia. The students started calling their protests "sit-downs" or "sit-ins." The prison sentences that often attended their activism did nothing to quell their ardor. "Dear Mom and Dad: I am writing this letter tonight from a cell in the Greensboro jail. I was arrested this afternoon when I went into a lily-white lunch room and sat down . . ." wrote a young Portsmouth woman who attended North Carolina A&T. Like a match on dry kindling, the sit-ins set aflame Negroes' smoldering, long-deferred dream of equality with a speed and intensity that took even the black community by surprise.

Hampton Institute was the first school outside of North Carolina to organize a sit-in. On the campus, many students had come into contact with one of the early icons of a mobilization that seemed to be gaining national momentum. Five years earlier, Rosa Parks, the Montgomery, Alabama, seamstress and NAACP member, refused to yield her seat on a city bus to a white man, galvanizing the bus boycott led by Martin Luther King Jr. and Ralph Abernathy. A ferocious backlash against Parks ensued: she received death threats, and both she and her husband, Raymond, were blacklisted from employment in Montgomery. The president of Hampton Institute reached out to Parks, offering her a job as a hostess at the university's faculty dining room, the Holly Tree Inn. Parks accepted, arriving on campus in 1957 and working at the restaurant into 1958.

When the sit-ins came to Hampton, Christine Mann was an eighteen-year-old Hampton Institute junior carrying a double course load. Her father had insisted that she earn a teaching certificate as a backup plan for her pursuit of a career in the sciences. Christine found herself captivated by the incipient activist movement, and despite carrying a full semester of courses in math and physics and extra classes in teacher education, she found time to join the protests, which eventually swelled into marches of more than seven hundred. Students walked across the Queen Street Bridge to downtown Hampton and converged on the lunch counters at Woolworth's and Wornom's, the local drug store. They quietly occupied the stores, some sitting at tables reading and working on homework assignments, until the owners shut down their establishments in the middle of the afternoon. The next month, five hundred students staged a peaceful protest through downtown Hampton. An outspoken group of thirteen movement leaders held a press conference with local newspapers. "We want to be treated as American citizens," they told the reporters. "If this means integration in all areas of life, then that is what we want."

Christine also decided to join the voter registration drives organized at Hampton, walking door-to-door in black neighborhoods along Hampton's Shell Road and Rip Rap Road, imploring black voters to register in time to make their voices heard in the November 1960 presidential showdown between the Republican, Vice President Richard Nixon, and the Democrat, Senator John F. Kennedy of Massachusetts.

Despite its unyielding advocacy of Negro economic empowerment, Hampton Institute's stance on integration had always been of the go-slow variety, wartime president Malcolm MacLean being a notable exception. Now, with a black president at the helm for the first time, even Hampton succumbed to the zeitgeist of the era. Dorothy Vaughan's eldest daughter, Ann, who had left Hampton Institute in 1957, returned in the fall of 1959 to finish her degree. The campus she came back to was alive, breathless even, with the possibility of significant and permanent social change. One rumor that spread like wildfire through the network of energized students—a rumor that seemed wholly improbable, but which took root until it was accepted as fact—was that the astronauts were contributing to the students' organizing activities. The astronauts represented everything that mainstream America held dear—*and they're with us*, the students marveled. The very *idea*, that those buzz-cut middle-American boys were standing, however surreptitiously, with the Negro student activists! The fact that the rumor couldn't be confirmed did nothing to dampen its power. At the beginning of a decade when everything was beginning to seem possible, nothing seemed impossible.

If anyone could bear witness to the long-term impact of persistent action, and also to the strength of the forces opposing change, it was Dorothy Vaughan. Virginia's governor, Lindsay Almond, capitulated in the fight over schools, reopening Norfolk, Charlottesville, and Front Royal schools in 1959 and inching toward integration: eighty-six black students in those districts now attended school with whites. In Prince Edward County, however, segregationists would not be moved: they defunded the entire county school system, including R. R. Moton in Farmville, rather than integrate. No municipality in all of America had ever taken such draconian action. As white parents herded their students into the new segregation academies, the most resourceful black families scrambled to salvage their children's educations by sending them to live with relatives around the state, some as far afield as North Carolina. Prince Edward's schools would remain closed from 1959 through 1964, five long and bitter years. Many of the affected children, known as the "Lost Generation," never made up the missing grades of education. Virginia, a state with one of the highest concentrations of scientific talent in the world, led the nation in denying education to its youth. Dorothy's friends and former Moton colleagues watched helplessly as their children's futures were sacrificed in the battle over the future of Virginia's public schools. Commenting on the situation in 1963, United States Attorney General Robert Kennedy said, "The only places on earth known not to provide free public education are Communist China, North Vietnam, Sarawak, Singapore, British Honduras—and Prince Edward County, Virginia."

Meanwhile, Langley moved in the opposite direction. When Dorothy Vaughan turned off the lights in the West Area Computing office for the last time, she and the remaining women in the segregated pool were dispatched to the four corners of the laboratory, finally catching up to colleagues who had already found permanent positions in an engineering group. Marjorie Peddrew and Isabelle Mann went to Gas Dynamics, Lorraine Satchell and Arminta Cooke joined Mary Jackson in the Supersonic Tunnels Branch, Hester Lovely and Daisy Alston left for the Twenty-Inch Hypersonic Jets Branch, Eunice Smith went to Ground Loads, and Pearl Bassette was assigned to the Eleven-Inch Hypersonic Tunnel.

As for the West Computers' erstwhile leader, Dorothy Vaughan found herself in a new seat in another brand-new building. In 1960, Langley had only just completed Building 1268, a West Side facility housing one of the most advanced computer complexes on the East Coast. Electronic computing had moved from the wings of aeronautical research to the main stage. Accordingly, Langley centralized its computing operations into a group named the Analysis and Computation Division, created to service all the center's research operations, as well as to provide computing to outside contractors. The ACD organization chart was a snapshot of two decades of change at Langley. Dorothy was reunited with many of her West Computers, but they now worked side by side with East Computing alumni like Sara Bullock and Barbara Weigel.

Perhaps more striking than the racial integration of the female mathematicians, which had been spreading organically throughout Langley for years, was the fact that a group focused on computing now employed increasing numbers of men. The function of computing had been promoted from an all-female service organization with minimal hardware requirements to a top-level division with an eight-figure operating budget; it was starting to look a lot more like a launchpad and a career path to ambitious young men. The room-sized machines were remaking the old models of aeronautical research; their ascendance marked the beginning of an era that promised to be even more momentous than the one ushered in by the flying machine. For better or worse, it also signaled the beginning of the end of computing as women's work.

Some of the older women at the center, the ones who still relied upon the mechanical calculators, were starting to look as if they were stranded on an island, separated from the mainland by a gulf that grew wider each year. The early 1960s were an inflection point in the history of computing, a dividing line between the time when computers were human and when they were inanimate, when a computing job was handed off to a room full of women sitting at desks topped with $500 mechanical calculating machines and when a computing job was processed by a room-sized computer that cost in excess of $1 million.

Dorothy Vaughan was keenly aware of that undulating invisible line that separated the past from the future. At fifty years old and many years into her second career, she reinvented herself as a computer programmer. Engineers still made the pilgrimage to her desk, asking for her help with their computing. Now, instead of assigning the task to one of her girls, Dorothy made a date with the IBM 704 computer that occupied the better part of an entire room in the basement of Building 1268, the room cooled to polar temperatures to keep the machine's vacuum tubes from overheating.

In the past, Dorothy would have set up the equations in a data sheet and walked one of her girls through the process of filling it out. At

ACD, it was her job to convert the engineers' equations into the computer's formula translation language—FORTRAN—by using a special machine to punch holes in 7⅜"×3¼" cards printed with an array of eighty columns, each column displaying the numbers 0 through 9, each space assigned a number, letter, or character. Once punched, each cream-colored card represented one set of FORTRAN instructions.

The longer or more complex the program, the more cards the programmer fed the computer. The machines tapped out at two thousand cards—two thousand lines of instructions. Even modest programs could require a tray of hundreds of the cards, which needed to be fed into the computer in the correct order. Woe to the klutz who dropped a box of cards on the floor. Some programmers tried to forestall disaster by taking a Magic Marker and painting a big diagonal swath on the top surface of a vertical stack of cards, a continuous line from the front corner on the first card to the opposite back corner of the final card, hoping that the tiny dot of color on each would provide the key to reassembling the fumbled cards into the correct order.

As powerful as ACD's computer was, however, the maestros of Project Mercury would require even more electronic horsepower for what was to come next. At the end of 1960, NASA purchased two IBM 7090s and installed them in a state-of-the-art facility in downtown Washington, DC, managed by the Goddard Space Flight Center, a Greenbelt, Maryland, NASA field center opened in 1959 to focus exclusively on space science. The agency set up a third computer, a slightly smaller IBM 709, in a data center in Bermuda. Together the three computers would monitor and analyze all aspects of the spaceflights, from launch to splashdown.

The planned suborbital flights presented a controlled set of challenges. Taking off from Cape Canaveral, Florida, and landing in the Atlantic at a spot approximately fifty miles from Turks and Caicos, the hurtling capsule would remain within communications range of Mission Control in Florida and the data centers in DC and Bermuda. Orbital flights—which sent the astronaut on one or more ninety-minute circuits around the globe, passing out of visual and radio contact with Mission Control, flying over unfriendly territory—upped the ante by a factor. Constant contact with the astronaut during every minute of every orbit was a prerequisite for the flight.

The task of building a worldwide network of tracking stations that would maintain two-way communication between the orbiting space-craft and Mission Control fell to Langley. Langley put all available re-sources behind the $80 million project in 1960, putting the final pieces in place just before December 1960, the originally scheduled date for the first suborbital mission. The Mercury tracking network in and of itself was a project whose scale and boldness rivaled that of the space missions it supported. The eighteen communications stations set up at measured intervals around the globe, including two set up on navy ships (one in the Atlantic Ocean, another in the Indian Ocean), used powerful satellite receivers to acquire the radio signal of the Mercury capsule as it passed overhead. Each station transmitted data on the craft's position and speed back to Mercury control, which bounced the data to the Goddard computers. The "CO3E" software program, developed by the Mission Analysis branch and programmed into the IBM computers, integrated all the equations of motion that described the spacecraft's trajectory, ingested the real-time data from the remote stations, and then projected the remaining path of the flight, includ-ing its final splashdown spot. The computers also sounded the alarm at the first sign of trouble; any deviation from the projected flight path, evidence of malfunction on board the capsule, or abnormal vital signs from the astronaut, which were also being monitored and transmitted to doctors on the ground, would send Mission Control into trouble-shooting mode.

The launch date for Project Mercury's first manned mission slipped into 1961, a year that announced itself as unpredictable from the start: on January 3, the United States cut diplomatic relations with Cuba, another step down the road in the Cold War with the Soviet Union. President Dwight Eisenhower, in his farewell speech in January 1961, railed against the United States' growing military-industrial complex. On March 6, 1961, President John F. Kennedy, newly inaugurated, an-nounced Executive Order 10925, ordering the federal government and its contractors to take "affirmative action" to ensure equal opportunity for all of their employees and applicants, regardless of race, creed, color, or national origin. Through it all, the Space Task Group, the Langley Research Group, the other NASA centers, and thousands of NASA contractors pressed forward on their aerodynamic, structural, materials, and component tests, closing in on a target launch date in May.

On April 12, 1961, Russian cosmonaut Yuri Gagarin became in one fell swoop the first human in space and the first human to orbit Earth. "We could have beaten them, we should have beaten them," Project Mercury flight director Chris Kraft recalled decades later. But unlike the disorientation, anxiety, and fear that Sputnik provoked, the agency absorbed the blow. It was painful, certainly, and embarrassing as well, but they turned the welter of emotion into renewed intensity for the mission, employing all of their talents and the principles of math, physics, and engineering to create a precise and thorough plan. Now they executed it with the knowledge that there was only one direction to move: forward.

It would take a total of 1.2 million tests, simulations, investigations, inspections, verifications, corroborations, experiments, checkouts, and dry runs just to send the first American into space, a precursor to achieving Project Mercury's goal of placing a man into orbit. Every mission involved the Mercury capsule, though the rockets—Scout, Redstone, and Atlas—varied. Mercury-Redstone 1, or "MR-1," the first mission to mate the Mercury capsule to the Redstone rocket, failed on the launchpad. MR-2, with Ham the chimpanzee as its passenger, overshot the landing spot by sixty miles and was nearly underwater when it was finally plucked from the ocean. Pulling back the curtain on three and a half years of work, NASA took the audacious step of deciding to broadcast the launch of Project Mercury's first manned mission—"Mercury-Redstone 3," carrying astronaut Alan Shepard—live. Forty-five million Americans would tune in to witness the ultimate success or failure of MR-3. When Shepard finally strapped into the disarmingly small capsule—just six feet in diameter and six feet, ten inches high—

and rode the Redstone candle into space, reaching an altitude of 116.5 miles above Earth, it was a resurrection for the United States and a much-needed dose of adrenaline for NASA.

The suborbital flight in the capsule Shepard christened Freedom 7 lasted only fifteen minutes and twenty-two seconds and covered 303 miles, just about the distance between Hampton, Virginia, and Charleston, West Virginia. Freedom 7 was a pale technical achievement compared to Yuri Gagarin's orbital flight the month before, but its success emboldened President Kennedy to pledge the country to a goal significantly more ambitious: a manned mission to the Moon.

"I believe that this nation should commit itself to achieving the goal, before this decade is out, of landing a man on the Moon and returning him safely to the Earth," President Kennedy said before a session of Congress, not three weeks after Shepard splashed down. Every NASA employee involved with the space program, still burning the midnight oil working on Project Mercury, broke out in a cold sweat. The agency hadn't yet achieved its mandate to place a human into orbit, and Kennedy already had them kicking up Moon dust?

It was a terrifying prospect—and the most exhilarating thing they had ever heard. Unspoken publicly until that moment, getting to the Moon, one of mankind's deepest and most enduring dreams, had long been the private dream of many at Langley as well. But with only one operational success under its belt and with six Mercury missions to go—with the orbital flight still on the drawing board—NASA's road to the Moon seemed unimaginably complex. The engineers estimated that the upcoming orbital flight, including the fully manned global tracking network, required a team of eighteen thousand people. The buildup to a lunar landing would demand many times more people than could be reasonably supported by Mother Langley.

The whispered rumors now gained currency: the Space Task Group's time in Hampton was coming to an end. The Langley employees, and the locals, campaigned with all their might to keep their brainchild from leaving home. Geography and politics had smiled on Virginia in 1915, when the NACA first went searching for its proving ground and aeronautical laboratory. As it had in the period leading up to World

War I, the federal government made a list of possible sites for the head-quarters for its space effort, looking for the right combination of climate, available land, and friendly politicians. In 1960, nine locations made the short list, but Virginia was not one of them. Due in no small part to the influence of powerful Texans, including now Vice President Lyndon Johnson, NASA decided to move the heart of its space program to Houston. Many of the Langley employees—the former NACA nuts, including Katherine Johnson—were going to have to make hard choices. They had come to love their home by the sea, from the abundant fresh seafood to the mild winters to the water that surroun· d the lonely finger of land that had become such a part of them. Soon, they knew, following the president's lead into space might mean choosing between the place that had given them a community and the passion for the work that gave their life meaning.

Over in Building 580 on Langley's East Side, Katherine's former colleagues Ted Skopinski, John Mayer, Carl Huss, and Harold Beck, who led the Mission Analysis branch within the rapidly growing Space Task Group, prepared for the move to Houston. Mary Shep Burton, Catherine T. Osgood, and Shirley Hunt Hinson, the math aides who ran the trajectory analysis software on the group's IBM 704, also decided to go. Unless more Langley women volunteered to make the move, the members of the branch worried that their new office "was going to be badly understaffed" just as the workload skyrocketed.

Katherine Johnson had been asked to transfer to Houston with the group, but her husband, Jim, wanted them to stay close to their families. Resisting Houston's call, not following the nerve center of the space program across the country, was difficult for Katherine and many of her Langley colleagues. It was "impractical" to recruit the mathematicians they needed in Virginia, so Mary Shep Burton and John Mayer went to Houston to recruit "five qualified young women" to come to Langley for training before setting up a permanent new computing pool in the under-construction "Manned Spacecraft Center." The move echoed the establishment of the first computing pool at Langley twenty-five years before.

The residents of Building 1244 might have been staying put in Hampton, but despite their concerns, much work remained for them on Project Mercury. Alan Shepard's flight was a triumph. MR-4, Virgil "Gus" Grissom's July 1961 suborbital flight, came and went in a flash.

NASA's first orbital mission, and the debut of the all-important tracking and communications network, shimmered in the near distance like a heat mirage. Katherine and Ted Skopinski had laid out the fundamentals of the orbital trajectory nearly two years earlier, in their important Azimuth Angle report, then handed off the responsibility for the calculation of the flight launch conditions to the IBM computers. Like Dorothy Vaughan, Katherine Johnson knew that the rest of her career would be defined by her ability to use the electronic computers to transcend human limits. But before she crossed completely to the world of electronic computing, Katherine Johnson would tackle one last, very important assignment, using the techniques and the tools that belonged to the human era of computing. Like her fellow West Virginian John Henry, the steel-driving man who faced off against the steam hammer, Katherine Johnson would soon be asked to match her wits against the prowess of the electronic computer.

CHAPTER TWENTY-ONE

Out of the Past, the Future

Sending a man into space was a damn tall order, but it was the part about returning him safely to Earth that kept Katherine Johnson and the rest of the space pilgrims awake at night. Each mission presented myriad pathways to disaster, starting with the notoriously temperamental Atlas rocket, a ninety-five-foot-high, 3.5-million-horsepower intercontinental ballistic missile that had been modified to propel the Mercury capsule into orbit. Two of the Atlas's last five sallies had ended in failure. One of them had surged into the sky before erupting into spectacular fireballs with the capsule still attached. That wasn't exactly a confidence builder for the man preparing to ride it into orbit, but it was the more powerful Atlas that would be required to accelerate the Mercury capsule to orbital velocity. The capsule itself was the most sophisticated tin can on the planet. The vehicle's oxygen and pressurization systems stood between the astronaut and the life-crushing vacuum of space. Those functions and more—every switch, every indicator, every gauge—had to be tested and retested for any whiff of possible failure. As the rocket blasted from the launchpad and accelerated into the sky toward maximum velocity, the aerodynamic pressure on the capsule also increased to a point known as "max Q." If the capsule wasn't strong enough to withstand the forces acting on it at max Q, it could simply explode. A Republican senator from Pennsylvania called the Mercury capsule-Atlas rocket pairing "a Rube Goldberg device on top of a plumber's nightmare."

Everything rested upon the brain busters' mastery of the laws of physics and mathematics. The mission was colossal in its scope, but it required both extreme precision and the utmost accuracy. A number transposed in calculating the launch azimuth, a significant digit too few in measuring the fully loaded weight of the capsule, a mistake in accounting for the rocket's speed and acceleration or the rotation of the Earth could cascade through the chain of dependencies, causing serious, perhaps catastrophic, consequences. So many ways to screw the pooch, and just one staggeringly complex, scrupulously modeled, endlessly rehearsed, indefatigably tested way to succeed.

Nobody, of course, understood this better than astronaut John Glenn. In 1957, the former Marine test pilot had campaigned fiercely—and unsuccessfully—to be the first of the Mercury Seven to navigate to the heavens. Now, NASA had picked Glenn for MA-6, the orbital flight that would cast the die on the space agency's future, and he was leaving nothing to chance. He pushed himself to his physical limit, running miles each day to stay fit, tag-teaming with fellow astronaut Scott Carpenter to practice water egress from the capsule in the Back River on Langley's East Side. With the experience of Alan Shepard and Gus Grissom as a guide, NASA physicians worried a little less about the health hazards Glenn might face while on board, since in the capsule he would be hooked up to wires like a lab rat, his every vital sign transmitted to and monitored by the doctors on the ground. The specter of human error was ever-present, of course, so Glenn worked the simulators and procedures trainers obsessively, putting himself through hundreds of simulated missions, honing his responses to every failure scenario the engineers could imagine.

As a seasoned test pilot, Glenn knew that the only way to remove all danger from the mission was to never leave Earth. The former Marine was the first pilot to average supersonic speeds in a transcontinental flight. From Project Mercury's outset, the NASA engineers had the delicate task of balancing the drive to get into space as quickly as possible with the risk they felt they could reasonably ask their human cargo to accept. Experience and analysis informed them that somewhere along this venturesome path they were certain to encounter unforeseen problems or run smack into the simple bad luck of statistics, that one time in a thousand when the worst-case scenario played out. What was within their control, however, they took pains to bulletproof, even if it meant stretching—then breaking—their timeline. Project Mercury's first orbital flight was originally slated to take place at the end of 1960, on President Eisenhower's watch. Additional testing and fine-tuning—a cooling system bug here, an oxygen delivery glitch there, the need to implement improvements based on previous unmanned and suborbital flights—conspired to push the date into the administration of incoming President Kennedy. NASA set July 1961 as the new date for the orbital flight, then rescheduled it for October, then to December. Finally the mission slipped into 1962.

While NASA appeared to be dithering on the ground, Russian cosmonaut Gherman Titov followed Yuri Gagarin's April 1961 triumph with a successful seventeen-orbit flight, nearly a *full day* in space, on October 6, 1961. American government officials, the press, and the public expressed their disappointment with the delays, many impugning the agency's judgment and competence. Even when the technical issues were in hand, the launch team had to contend with the weather. A long stretch of low, overcast skies at Cape Canaveral put the kibosh on two more scheduled launches, on January 20 and February 12, 1962. Finally, the Space Task Group affixed February 20, 1962, as John Glenn's debut.

The incessant delays and high stakes would have caused most individuals to lose their focus, but John Glenn gave even-tempered, optimistic interviews to the impatient press and busied himself by keeping his mind and body in peak condition. Three days prior to the most significant date of his life, Glenn went through a final simulation, carrying out a full checkout of his flight plan. Before commending himself to his destiny, however, the astronaut implored the engineers to execute one more check: a review of the orbital trajectory that had been generated by the IBM 7090 computer.

Many of the operational aspects of John Glenn's upcoming flight had been refined by testing during the years following Sputnik, and the knowledge and experience gleaned during the early days were consolidated during the execution of the suborbital flights. The recovery team confidently manned their stations around the globe, ready to haul the astronaut and his capsule out of the water. NASA put considerable effort into building redundancies and fail-safes into the network of IBM computers and the eighteen-station Mercury tracking network.

The astronauts, by background and by nature, resisted the computers and their ghostly intellects. In a test flight, a pilot staked his reputation and his life on his ability to exercise total, direct, and constant control over the plane. A tiny error in judgment or a speck of delay in deciding on a course of action might mean the difference between safety and calamity. In a plane, at least, it was the pilot's call; the "fly-by-wire" setup of the Mercury missions, where the craft and its controls were tethered via radio communication to the whirring electronic computers on the ground, pushed the hands-on astronauts out of their comfort zone. Every engineer and mathematician had a story of double-checking the machines' data only to find errors. What if the computer lost power or seized up and stopped working during the flight? That too was something that happened often enough to give the entire team pause.

The human computers crunching all of those numbers—now *that* the astronauts understood. The women mathematicians dominated their mechanical calculators the same way the test pilots dominated their mechanical planes. The numbers went into the machines one at a time, came out one at a time, and were stored on a piece of paper for anyone to see. Most importantly, the figures flowed in and out of the mind of a real person, someone who could be reasoned with, questioned, challenged, looked in the eye if necessary. The process of arriving at a final result was tried and true, and completely transparent.

Spaceship-flying computers might be the future, but it didn't mean John Glenn had to trust them. He did, however, trust the brainy fellas who controlled the computers. And the brainy fellas who controlled the computers trusted *their* computer, Katherine Johnson. It was as simple as eighth-grade math: by the transitive property of equality, therefore,

John Glenn trusted Katherine Johnson. The message got through to John Mayer or Ted Skopinski, who relayed it to Al Hamer or Alton Mayo, who delivered it to the person it was intended for.

"Get the girl to check the numbers," said the astronaut. If she says the numbers are good, he told them, I'm ready to go.

The space age and television were coming into their own at the same time. NASA was acutely aware that the task before them wasn't only about making history but also about making a myth, adding a gripping new chapter to the American narrative that worshipped hard work, ingenuity, and the triumph of democracy. At the Cape, a behind-the-scenes camera captured extensive footage of the astronaut as he walked through each station of the trip he had already taken hundreds of times in NASA simulators, fodder for a documentary to be released later in the year. The agency sent a film crew to each of the remote tracking stations, recording the communications teams as they completed their preflight checkouts. And the footage that showed the second-by-second drama in Mission Control—white guys in white shirts and skinny black ties wearing headphones, facing forward at long desks outfitted with communications consoles, mesmerized by the enormous electronic map of the world on the wall in front of them—created the enduring image of the engineer at work.

Meanwhile, away from the front lines, out of sight of the cameras, the black employees, whose numbers had been growing at Langley and all the NASA centers since the end of World War II, busily calculated numbers, ran simulations, wrote reports, and dreamed of space travel alongside their white counterparts, as curious as any other brain buster about what humanity might find once it had ventured far from its spherical island, and just as doggedly pushing for answers to their inquiries. At the Lewis Research Center in Ohio, a black scientist named Dudley McConnell was among the researchers working on aerodynamic heating, one of the most serious challenges facing the astronauts as they reentered Earth's atmosphere and plummeted toward the ocean. Annie Easley, who had joined the Lewis Laboratory in 1955, was staffed on Project Centaur, developing a rocket stage that was ultimately used in the Atlas. At the Goddard Space Flight Center in Maryland, which was charged with the operation of the two IBM 7090s that would track the spaceship and relay information to Mission Control, a Howard University graduate named Melba Roy oversaw a section of programmers working on trajectories.

Also at Goddard was Dorothy Hoover, embarking on the third (or fourth, or maybe fifth) act of her career. Following her graduate work at the University of Michigan, Hoover had worked at the Weather Bureau for three years. Perhaps nostalgic for the agency that had boosted her mathematical career, she transferred to Goddard in 1959, the only one of the centers that had been created organically out of NASA. Her career advance had continued; she now held a senior ranking of GS-13. While her colleagues at Langley put their minds to work on the engineering project of the century, Dorothy Hoover folded herself back into the theoretical work she loved, continuing her publication record with a coauthored book on computational physics.

It was at Langley where the progress of the last two decades was most evident. At the Transonic Dynamics Tunnel, Thomas Byrdsong got a head start on the long road to the Moon by testing a model of the Saturn rocket, a launch vehicle the size of a redwood tree. Engineer Jim Williams, still on the team with John D. "Jaybird" Bird, was already helping to work toward President Kennedy's pledge of a Moon landing. The division would be associated with lunar orbit rendezvous, one of the most ingenious and elegant solutions to the challenge of propelling extraordinarily heavy objects on the several-hundred-thousand-mile journey to the Moon and back.

West Computing no longer existed as a physical space, but its alumni pushed their minds and hands in the service of the space program—though in Dorothy Vaughan's case, it was an indirect effort. The computer minders of the two IBM 7090s being used to track the flight were ensconced at Goddard, and much of the analysis was being done in the Space Task Group's Mission Planning Analysis Division. The women and men in ACD were as busy as ever, however. Dorothy's hunch that those who knew how to program the devices wouldn't want for work was a correct one. Though she wasn't on the front lines of the programming that was being done for Project Mercury, she did have a hand in the calculations that were used in Project Scout, a solid-fuel rocket that Langley tested at the Wallops Island facility. She had even been making trips up to the test range for work. The Scout rocket had been an important part of laying the groundwork for the manned spaceflight efforts. Engineers used it to take a "dummy" astronaut, weighing as much as a real astronaut, for a four-orbit flight in November 1961.

Other West Computers had a closer view. Miriam Mann worked for Jim Williams, running the numbers for the "rendezvous" research that would allow two vehicles to dock while in space. At the Four-foot SPT, Mary Jackson conducted tests of the Apollo capsule and other components, honing their fitness for the portion of the journey that would take place in the supersonic speed regime. That work would earn her an Apollo Team Achievement Award. Sue Wilder was rolling up her sleeves among the "mad scientists" of Langley's Magnetoplasmadynamics (MPD) Branch, her work also concerned with the physics of a vehicle reentering the atmosphere.

But because of her close working relationship with the pioneers of the Space Task Group, it was Katherine Johnson who found herself in a position to make the most immediate contribution to the pageant that was about to begin in Florida. The broader implication of her role as a black woman in a still-segregated country, helping to light the fuse that would propel that country to achieve one of its greatest ambitions, was a topic that would occupy her mind for the rest of her life. But with the final countdown in sight, that was a matter for the future. Right now, she was a mathematician, an American citizen whose greatest talents had been recognized, and who was about to offer those talents in the service of her country. Katherine Johnson had always been a great believer in progress, and in February 1962, once again, she became its symbol.

When the phone call came in, forty-three-year-old Katherine was at her desk in Building 1244. She overheard the call with the engineer who picked it up, just as she had overhead the conversation between Dorothy Vaughan and the engineer in 1953, the request that sent her to the Flight Research Division two weeks after she arrived at Langley. She knew she was the "girl" being discussed in the phone conversation. She had seen the astronauts around the building, of course; they had spent many hours in the hangar downstairs, preparing for their missions on a simulation machine called the Procedures Trainer. Some of their briefings with the brainy fellas had happened upstairs, though she was not invited to attend those meetings. That John Glenn didn't know, or didn't remember, her name didn't matter; what did matter, as far as he was concerned—as far as *she* was concerned—was that she was the right person for the job.

Many years later, Katherine Johnson would say it was just luck that of all the computers being sent to engineering groups, she was the one sent to the Flight Research Division to work with the core of the team staffed on an adventure that hadn't yet been conceived. But simple luck is the random birthright of the hapless. When seasoned by the subtleties of accident, harmony, favor, wisdom, and inevitability, luck takes on the cast of serendipity. Serendipity happens when a well-trained mind looking for one thing encounters something else: the unexpected. It comes from being in a position to seize opportunity from the happy marriage of time, place, and chance. It was serendipity that called her in the countdown to John Glenn's flight.

In the final section of the Azimuth Angle research report she completed in 1959, Katherine had marched through the calculations for two different sample orbits, one following an eastward launch and the other a westward, as Glenn was scheduled to fly. Once she had worked out the math for the test scenarios on her calculating machine, substituting the hypothetical numbers for variables in the system of equations, the Mission Planning and Analysis Division within the Space Task Group took her math and programmed it into their IBM 704. Using the same hypothetical numbers, they ran the program on the electronic computer, to the pleasing end that there was "very good agreement" between the IBM's output and Katherine's calculations. The work she had done in 1959, double-checking the IBM's numbers, was a dress rehearsal—a simulation, like the ones John Glenn had been carrying out—for the task that would be laid on her desk on the defining day of her career.

When the Space Task Group upgraded their IBM 704 to the more powerful IBM 7090s, the trajectory equations were programmed into those machines, along with all the other programs required to guide and control the rocket and capsule and compare the vital signs of the flight at every moment to the flight plan programmed into the computer. During the launch phase of the mission, a computer in the Atlas rocket, programmed with the launch coordinates, communicated with Mission Control. If the rocket misfired and was on track to inject the capsule into an incorrect orbit, the flight controllers could decide to abort the mission—a go–no go moment—automatically detaching the capsule from its rocket and sending it off into the sea in a mangled suborbital trajectory.

Once the capsule climbed through the launch window, separated from the Atlas, and settled into a successful orbit, it established communication links with the ground stations. As the craft flew overhead, it telemetered a torrent of data to the closest tracking station, everything from its speed and altitude to its fuel level and the astronaut's heart rate. The tracking stations captured the signals with their sixty-four-foot receiving dishes, then relayed this data plus voice communications through a jumble of submarine cables, landlines, and radio waves to the computer center at Goddard. The IBM machines used the inputs they received to make calculations based on the orbit determination programs. Via high-speed data lines—a blazing 1 kilobyte per second—Goddard sent Mission Control real-time information on the spaceship's current position. There, on the front wall of the room that served as NASA's nerve center, was a huge lighted map of the world. On the map were inscribed sine wave–shaped tracks, one for each orbit. Hovering over the map was a little cutout of a Mercury capsule, suspended on a wire. As tracking data from the spaceship filtered into Mission Control, the toy capsule moved along the orbit grooves on the map too, a puppet controlled by its master in the sky. The capsule's signal bounced from one tracking station to the next as the orbit proceeded, like a very fast and expensive game of telephone, constantly communicating its position and status. *He's passing over Nigeria! He's just about to reach Australia!* The crude setup seemed like a miracle:

looking at the puppet ship, they could actually "see" the spaceship as it made its rounds.

The Goddard computers also sent the flight controllers their projection of the remainder of the voyage. Where was the capsule compared to where they had calculated it to be at the given time? Was it too high, too low, too fast, too slow? The output included a constantly updated time for retrofire, the moment when the capsule's rockets had to be fired in order to initiate its descent back to Earth. Retrofiring too soon or too late would bring the unlucky astronaut back down far afield of his navy rescuers.

The engineers had actually taken the IBM 7090 and the orbital equations for a test drive on two prior occasions: once for Mercury-Atlas 4, an orbital flight using a mechanized astronaut "dummy" as a passenger, and then with the trained chimpanzee Enos at the controls of MA-5. Enos' flight was ultimately successful, but it faced computer glitches and communications dropouts (in addition to more serious problems with the capsule's cooling system and a faulty electrical wire). To mention that the stakes increased dramatically with a person on board was an understatement (if disaster did befall John Glenn, one secret military document proposed blaming it on the Cubans, using it as an excuse to overthrow Fidel Castro). Katherine Johnson, suffice it to say, was very nervous about the momentous task she had been handed.

For the entire project to succeed, each individual part of the mission—the hardware, the software, and the human—had to function according to plan. A breakdown would be immediate and potentially tragic, and broadcast live on television. But Katherine Johnson, like John Glenn, was not prone to panic. Like him, she had already gone through a simulation of the job in front of her. The moment that had arrived, despite the time pressure and the frenzy of activity surrounding her, felt somehow inevitable. Katherine Johnson's life had always seemed to be guided by a kind of providence, one that was unseen by others and not fully understood by her, perhaps, but obeyed by all who knew her, the way one obeys the laws of physics.

•   •   •

Katherine organized herself immediately at her desk, growing phone-book-thick stacks of data sheets a number at a time, blocking out everything except the labyrinth of trajectory equations. Instead of sending her numbers to be checked by the computer, Katherine now worked in reverse, running the same simulation inputs that the computer received through her calculator, hoping that there would be "very good agreement" between her answers and the 7090s', just as had been the case when she originally ran the numbers for the Azimuth Angle report. She worked through every minute of what was programmed to be a three-orbit mission, coming up with numbers for eleven different output variables, each computed to eight significant digits. It took a day and a half of watching the tiny digits pile up: eye-numbing, disorienting work. At the end of the task, every number in the stack of papers she produced matched the computer's output; the computer's wit matched hers. The pressure might have buckled a lesser individual, but no one was more up to the task than Katherine Johnson.

February 20 dawned with clearing skies. No one who witnessed the events of the day would ever forget them. One hundred thirty-five million people, an audience of unprecedented size, tuned in to watch the spectacle as it unfolded on live television. Many Langley folks joined the Space Task Group down at Cape Canaveral to see the flight in person. Katherine sat tight in the office, watching the transmission on television.

At 9:47 a.m. EST, the Atlas rocket boosted Friendship 7 into orbit like a champion archer hitting a bull's-eye. The insertion was so good that the ground controllers cleared Glenn for seven orbits. But then, during the first orbit, the capsule's automatic control system began to act up, causing the capsule to pull back and forth like a badly aligned car. The problem was relatively minor; Glenn smoothed it out by switching the system to manual, keeping the capsule in its correct position the same way he would have flown a plane. At the end of the second orbit, an indicator in the capsule suggested that the all-important heat shield was loose. Without that firewall, there was nothing standing between the astronaut and the 3,000-degree Fahrenheit temperatures—almost as hot as the surface of the Sun—that would build up around the capsule as it passed back through the atmosphere. From Mission Control came an executive decision: at the end of the third orbit, after the retrorockets were to be fired, Glenn was to keep the rocket pack attached to the craft rather than jettisoning it as was standard procedure. The retropack, it was hoped, would keep the potentially loose heat shield in place.

At four hours and thirty-three minutes into the flight, the retrorockets fired. John Glenn adjusted the capsule to the correct reentry position and prepared himself for the worst. As the spaceship decelerated and pulled out of its orbit, heading down, down, down, it passed through several minutes of communications blackout. There was nothing the Mission Control engineers could do, other than offer silent prayers, until the capsule came back into contact. Fourteen minutes after retrofire, Glenn's voice suddenly reappeared, sounding shockingly calm for a man who just minutes before was preparing himself to die in a flying funeral pyre. Victory was nearly in hand! He continued his descent, with the computer predicting a perfect landing. When he finally splashed down, he was off by forty miles, only because of an incorrect estimate in the capsule's reentry weight. Otherwise, both computers, electronic and human, had performed like a dream. Twenty-one minutes after landing, the USS *Noa* scooped the astronaut out of the water.

John Glenn had saved America's pride! That he'd had to stare death in the face to do so only increased the power of the myth that was created that day. An audience with the president, a ticker-tape parade in New York, seventy-two-point newspaper headlines from Maine to Moscow. America couldn't get enough of its latest hero. Even the Negro press cheered Glenn's accomplishment. "All of us are happy to call him our Ace of Space," wrote a columnist in the *Pittsburgh Courier*.

Nowhere, perhaps, was the hero's welcome as warm as in Hampton Roads. Thirty thousand local residents turned out on a blustery day in mid-March to fete the men they had adopted as hometown heroes. Not since the end of the last war had Hampton seen such an exuberant celebration. Glenn rode in the lead vehicle of the fifty-car parade carrying the Mercury astronauts and their families and the top leadership of NASA. The motorcade departed from Langley Air Force Base and traced a twenty-two-mile route through Hampton and Newport News: along the shipyard, over the Twenty-Fifth Street Bridge, down Military Highway, with throngs standing on the sides of every thoroughfare. The procession passed by Hampton Institute, cheered on by Katherine Johnson's daughter Joylette and Dorothy Vaughan's son, Kenneth. Tiny Christine Darden stood on tiptoe to see over the exuberant crowds.

The parade ended at Darling Stadium, the namesake of the oyster magnate whose creative entrepreneurship had brokered the land deal with the federal government for the Langley laboratory a half century before. Glenn ascended to the podium, grinning broadly as he stood behind a sign reading SPACETOWN, USA. The people of Hampton and Newport News beamed with pride. With the heart of the space program shipping out for Houston, the celebration was tinged with melancholy, but the cities of the Virginia Peninsula were determined to commemorate their legacy as the birthplace of the future. The city of Hampton changed its official seal to depict a crab holding a Mercury capsule in its claw, adopting the motto *E Praeteritis Futura*: Out of the past, the future. Military Highway, the town's main drag since Hampton's days as a war boomtown, got a new name: Mercury Boulevard.

John Glenn was a bona fide hero, but he wasn't the only one being cheered. Word of Katherine Johnson's role in Glenn's successful mission began making the rounds in the black community, first locally, then farther afield. On March 10, 1962, a glamorous Katherine Johnson, bedecked in pearls and an elegant suit that would have made Jackie Kennedy proud, smiled from the front page of the *Pittsburgh Courier*. "Her name . . . in case you haven't already guessed it . . . is Katherine Johnson: mother, wife, career woman"! (Below the feature on Katherine Johnson, another headline inquired: "Why No Negro Astronauts?") The newspaper recounted the lady mathematician's background and accomplishments with pride, detailing the report that sent Glenn's rocket cone whizzing through the sky. Katherine accepted the recognition graciously: all in a day's work.

She and some of the engineers turned out for the parade, enjoying the celebration, allowing themselves, perhaps, just a sliver of pride in having been a part of such an achievement. They watched for a while but didn't tarry long. It was fine to celebrate past accomplishments, but there was nothing more exhilarating than getting back to work on the next thing.

CHAPTER TWENTY-TWO

America Is for Everybody

America Is for Everybody," proclaimed the US Department of Labor brochure that landed on Katherine Johnson's desk in May 1963. On the cover, a black boy of eight or nine, barefoot and dressed in a striped short-sleeved shirt and worn dungarees, sat on the ties of a dusty railroad track, his apparent circumstances and open-faced glower a rebuke to the promise of the title. Inside, President Kennedy and Vice President Johnson waxed poetic in statements about the Negro's epic hundred-year journey up from slavery. Photos of black employees who "occupied positions of responsibility" at NASA, all of them involved with the space program, accompanied the text. At NASA's High-Speed Flight Research Center on Edwards Air Force Base—the place where pilot Chuck Yeager first cracked through the sound barrier in 1947—engineer John Perry manned an X-15 simulator. Mathematicians Ernie Hairston and Paul Williams conferred on "orbital elements, capsule position, and impact points" at Goddard. One picture showed Katherine Johnson sitting at her desk at 1244, pencil in hand, "analyzing lunar trajectories and computing trip time to the Moon and return to Earth by a space vehicle." The document, created by the Labor Department to commemorate the centennial of the Emancipation Proclamation, certainly also served as another propaganda tool for the US government to improve its image on racial relations. What could provide greater evidence of America's growing commitment to democracy for all than the hard-at-work photos of Katherine Johnson and the seven other NASA employees—all men—profiled in the booklet? The resonances and dissonances of the images in the book were sharpest there at Langley, ten miles from the point where African feet first stepped ashore in English North America in 1619, less than that from the sprawling oak tree where Negroes of the Virginia Peninsula convened for the first southern reading of the Emancipation Proclamation. In a place with deep and binding tethers to the past, Katherine Johnson, a black woman, was midwifing the future.

Katherine wasn't the only one working with purpose in 1963: so was the grand old man of the civil rights movement, A. Philip Randolph. As Gordon Cooper brought Project Mercury to a successful conclusion in 1963 with a twenty-two-orbit flight, Randolph made plans for another march on Washington. Unlike 1941's ghost rally—the march that never happened, the impetus for Roosevelt's Executive Order 8802 opening federal jobs to Negro employees—Randolph was going to see this one through. Working with activist Bayard Rustin, allied with Martin Luther King Jr., Randolph brought together a group that would come to be seen as the pantheon of leaders from the most energetic phase of the civil rights movement, including Dorothy Height, John Lewis, Daisy Bates, and Roy Wilkins.

The March on Washington for Jobs and Freedom was held August 28, 1963, attracting as many as three hundred thousand people to the nation's capital. Mahalia Jackson, Bob Dylan, and Joan Baez all took to the stage, musical witnesses to the idealism, hope, and persistence of a movement that drew its strength from its desire to force America to live up to its founding principles. Marian Anderson sang "He's Got the Whole World in His Hands," her rich contralto flowing like honey over the massive gathering, mesmerizing the crowd just as she had Dorothy Vaughan and her young children at Hampton Institute in 1946. The morning schedule was interrupted with news that was cause for grief, reflection, and a kind of sober hope: ninety-five-year-old W. E. B. Du Bois had died early that morning in Ghana, the country he had adopted as his home after battling the State Department to keep his American passport. From his birth in 1868, Du Bois' life bridged the years of Reconstruction and the twentieth-century movement. Mary MacLeod Bethune, A. Philip Randolph, Charles Hamilton Houston, and so many more had built their life's work on Du Bois' foundation. Now, it was time to pass the torch again.

Thirty-four-year-old Dr. Martin Luther King Jr. ascended the platform to address the crowd, easing into his prepared remarks. Then Mahalia Jackson, sitting behind King on the podium, shouted out "Tell them about the dream, Martin!" King pushed his written speech aside, gripped the lectern with both hands, and gave his country seventeen of the most memorable minutes in its history. There was America before King's "I Have a Dream" speech, and there was America after; King's message would ever after remind all the citizens of the nation that the Negro dream and the American dream were one and the same. Backstage at the march at the end of the day, seventy-four-year-old A. Philip Randolph was speechless, the tears in his eyes the only adequate expression of what it meant to him to see this day come to pass.

It's doubtful that Randolph ever knew how directly the 1941 March on Washington influenced the group of people whose work was the lifeblood of America's space program, but back at the Langley Research Center, a handful of black women could testify to the connection. "Dear Mrs. Vaughan: Our records indicate that you recently completed 20 years of Federal Government service," wrote Langley's director, Floyd Thompson, in the summer of 1963. A gold-and-enamel lapel pin adorned with a ruby was to be bestowed upon her at the center's annual awards ceremony, which recognized employees hitting milestones of service with the center.

Against the odds and contrary to their expectation when they first walked through the doors at Langley, the women of West Computing had managed to turn their wartime service into lasting and meaningful careers. By the standards of their parents and grandparents, and compared to many of their contemporaries, they had reached the mountaintop. Despite their progress, however, there was still work left to do at Langley. Breaking out of the paraprofessional status of computer or math aide presented a challenge for all women, more so for the black women. Of all the black employees working in research at Langley in the early 1960s, there were still only five categorized as engineers and sixteen with the title of mathematician. In a letter to NASA administrator James Webb, Langley's director, Floyd Thompson, lamented that "very few Negroes" were applying for open science and engineering positions at the laboratory. "There is no doubt that one of the reasons they do not apply is that they do not believe that the living conditions in the area would be favorable to them because the Langley Research Center, which is completely integrated, is situated in a community where social segregation based on color is still practiced to a certain extent."

With an all-out need for technical expertise to feed the space program, and with continued pressure from the federal government to remove race-based barriers in its organization, Langley redoubled its recruitment efforts, casting a wider net at black colleges that had turned out generations of black math and science graduates, like Hampton Institute, Virginia State University in Petersburg and its branch campus in Norfolk, North Carolina A&T, and other schools in nearby states. Many in the generation of Negro students who came of age in a decade defined by *Brown v. Board of Education* and Sputnik—the ones who in the future would be known as the civil rights generation—were drawn into the engineering profession for the "economic and social mobility" that was the result of the national demand for technical skills. Most of them were southerners; for them, there was no need to adjust to living conditions that they had known all their lives. In the mid-1960s, with "dreams of working at NASA," greater numbers of black college students found their way to Langley. Many of them were taken under the protective wing of Mary Jackson, who, like an ambassador, helped the recruits find places to live and settle into their jobs. She and Levi opened their house to them if they needed a home-cooked meal, or simply a place to go when they felt homesick. Mary and the other black employees at Langley tended the new recruits as carefully and lovingly as if they were a garden. Unlike the women who started in West Computing after years of teaching, the new generation was coming to research early in their careers—early enough that they'd have time to stretch out and see where their talent might take them.

One Sunday in 1967, after services at Carver Presbyterian Church, Katherine Johnson spied a new face in the crowd, a young woman who had come to the church with her husband and two young daughters. Always among the first to welcome new parishioners, Katherine strode forward and offered her hand in welcome. "I'm Katherine Johnson," she said. "Yes, I know," said Christine Mann Darden, now married, "you're Joylette's mother." Though she hadn't seen her for many years, Christine had met Katherine once at an AKA sorority barbecue hosted at the Johnsons' home.

Christine hadn't set out to find a job in aeronautical research. In the spring of 1967, as she ticked off the days to graduation from her master's degree program at Virginia State University, she had visited the school's placement office in order to apply for professor positions at Hampton Institute and Norfolk State. "We wish you'd been here yesterday," said the placement officer, "because NASA was interviewing." The woman handed Christine an application for federal employment. "Fill this out and bring it back tomorrow."

Christine's application was received enthusiastically; a follow-up phone call from Langley's Personnel Department turned into a day of interviews and then a job offer as a data analyst in the Reentry Physics branch. She reported first to former East Computer Ruby Rainey, then to former West Computer Sue Wilder. Christine commuted from Portsmouth for a short time before moving her family to Hampton after Sue Wilder tipped her off on a house for rent in her neighborhood. Once on the peninsula, Christine saw Katherine Johnson and many of the other former West Computers on a regular basis. They hosted card parties and invited her along, introducing her to the black community in Hampton and Newport News. Despite spending four years at Hampton Institute, she had rarely left the campus and came to the city as a virtual stranger. The network of older women helped her settle quickly in her new town.

Katherine Johnson invited Christine to join the choir at Carver. If

Katherine and Eunice Smith took time off work to sing for a funeral at the church, Christine often came along. Christine also ran into Katherine at local sorority activities. For many years, Katherine and Eunice Smith traded off serving as president and vice president of AKA's Newport News chapter, overseeing a busy schedule of events like the organization's annual picnic and the many scholarship fund-raisers that were a core part of the sorority's mission. Katherine Johnson was involved in so many civic and social associations—the Peninsula League of Women, which hosted an annual debutante ball for young black women; the Altruist Club, a middle-class social organization—that folks came to expect to see her broad smile and firm handshake wherever the professional set of the black community gathered. Even the brainy fellas in the office at 1244 knew that when the CIAA tournament came to town—the premier basketball event for black colleges—Katherine's desk would be empty, as she never missed her annual courtside date with Eunice Smith, the two rabid hoops fans having the time of their lives cheering for their favorites.

Christine Darden and Katherine Johnson got to know each other well outside the office, but they never had the chance to work together. Christine visited Katherine in her office a couple of times, but it would be years before Christine knew more about the full scope of her friend's mother's work. The press surrounding Katherine Johnson's role in John Glenn's flight had made her something of a celebrity in the local community and among the small national network of black engineers and scientists, but she remained modest about her work. "Well, I'm just doing my job," she would say, implying *and I'm assuming that you are, too.*

Of course, while Katherine took the accolades in stride, she never took the work for granted. Not a morning dawned that she didn't wake up eager to get to the office. The passion that she had for her job was a gift, one that few people ever experienced. That *did* make her special, she understood, and it bonded her to the engineers at work as strongly as social and charity activities bonded her to the women in her sorority. Together they shared the secret language of pericynthion altitudes and orbital planes and lunar equators. They experienced the indescribable joy of seeing their endeavors coalesce with those of the hundreds of thousands of other people now involved in the space program, the collective effort so much greater than the sum of the individual parts that it began to feel like a separate being. They also grieved together when all their best-laid plans were destroyed in the February 1967 electrical fire on board the Apollo 1 command module, which was on the launchpad in Cape Canaveral for testing. Fire flashed through the interior of the craft, and the three astronauts inside—Ed White, Roger Chaffee, and Gus Grissom of the Mercury Seven—perished instantly.

The tragic end of Apollo 1 shook NASA to its core. The astronauts weren't hundreds of thousands of miles away when the accident happened; they were on the ground, within feet of the ground crew and the engineers, and yet they still died. The road to the stars was a rough one, and the Apollo team needed no reminder of the risk. They redesigned the spacecraft, fixing flaws that had been exposed by the disaster and redoubling their focus on every possible detail of the next nine missions, each a step in the stairway to the Moon. The ascent to the Moon landing was predicated on the belief that each cell in the body of the space program was both individually superb and seamlessly connected to the cells around it.

Two vehicles and 238,900 miles: three days there and three days back. Twenty-one hours on the surface of the Moon for two astronauts in the lunar lander, while the service module circled the heavenly body in a parking orbit. Katherine knew better than anyone that if the trajectory of the parked service module was even slightly off, when the astronauts ended their lunar exploration and piloted their space buggy back up from the Moon's surface, the two vehicles might not meet up. The command service module was the astronauts' bus—their only bus—back to Earth: the lander would ferry the astronauts to the waiting service module and then be discarded. If the two vehicles' orbits didn't coincide, the two in the lander would be stranded forever in the vacuum of space.

The leadership of the Space Task Group set a risk standard of "three nines"—0.999, a criterion requiring that every aspect of the program be projected to a 99.9 percent success rate, or one failure for every thousand incidences. The astronauts, former test pilots and combat veterans accustomed to riding with the shadow of death on every flight, put themselves in NASA's hands. They were prepared to give their lives to the mission, just as they had been prepared to give their lives as pilots, but they would just as soon trust that the brain busters had done their math and that by the rule of three nines, their unprecedented flight to the Moon would be less risky in fact than a Sunday drive in their Corvettes.

Katherine Johnson, for her part, was determined to make this happen. She arrived at the office early, went home in the late afternoon to check on her daughters, and then came back in the evening, maintaining a schedule of fourteen- or sixteen-hour days. She and engineer Al Hamer collaborated on four reports between 1963 and 1969, some of them written to work out the all-important lunar orbits, others asking the question, What if? What if the computers went out? What if there was an electrical failure on board the spacecraft and the astronauts needed to navigate back home by the stars, like the mariners of a simpler age? As the years of the 1960s slipped by, it seemed that Katherine was increasingly at the office late at night, the hours passing like minutes as she and Hamer refined calculations and made rough drafts of diagrams for their reports.

One morning on the way to work, Katherine literally fell asleep at the wheel, waking up shaken but unhurt by the side of the road. She was so absorbed with the problem of keeping the astronauts safe on their round trip to the Moon that she was making herself vulnerable to the most mundane kind of risk. She had to keep pushing, however. Each year, NASA made progress toward converting the theoretical concepts of how to reach the Moon into operating practice. Each mission closed the distance, brought the fruit closer to their grasp. But this last step, with its complicated dance between the Moon and the lander and the waiting command module, was the most complicated. Katherine Johnson had given her best to her part of the grand puzzle, of that she was sure. The day was soon coming when the world would see if her best, if the brainy fellas' best, if NASA's best, was good enough.

CHAPTER TWENTY-THREE

To Boldly Go

In July 1969, a hundred or so black women crowded into a room, their attention commanded by the sounds and grainy images issuing forth from a small black-and-white television. The flickering light of the TV illuminated the women's faces, the history of their country written in the great diversity of their features and hair and skin color, which ranged from near-ivory to almost-ebony, hues of beige and coffee and cocoa and topaz filling in between. Some of the women were approaching their golden years, the passage of time and experience etched in their faces and bearing. Others were in the full bloom of youth, their eyes like diamonds, reflecting a bright future. They convened around the shared purpose of the advancement of women like them, and to use their collective talents for the betterment of their community. From up and down the East Coast, and even farther afield, they had come together for the weekend, though the time they shared in each other's company would forge lifetime friendships.

Their presence at the conclave tipped them as members of the top echelon of the race, though in fact many of them were the daughters and granddaughters of the janitresses and washerwomen and domestic servants whose backbreaking work had funded thousands of college educations and bankrolled down payments on homes, who supported

America's great economic pyramid even as it pinned them in place with its weight. They, the legacy of those women, had spent their lives in varying degrees of distance from their country's great pageant, standing on the side of the stage, even though there was virtually no aspect of their lives that had not been touched by those big sweeps, no part of the grand story that did not include them in some way.

Throughout the day, as the women conducted their meeting, their interest in the marathon drama unfolding on the television had waxed and waned, groups of them perching in front of the screen to get the latest updates before heading off to focus on the day's agenda. But as the day grew long, more of them were drawn in, gazing into the television—into the void of space, into their own hearts, trying to make some meaning of what they were watching. The women who watched joined with their fellow Americans in a moment of consonance, the roulette of emotions in the room—pride, elation, impatience, awe, resentment, patriotism, suspense, fear—replayed in differing mixtures in living rooms and meeting places around the United States. In fact, the unprecedented episode they witnessed on that Saturday night they shared with a total of six hundred million people around the world: all standing in front of the same window, all observing the same thing at the same time.

Out of that global audience, four hundred thousand NASA employees, contractors, and military support watched with particular interest, seeing in the craft that approached the Moon the measure of a screw, the blueprint of a hatch, the filament in a circuit, the fulfillment of a promise made by a president who hadn't lived to see it carried out. They dotted the globe, those who had worked on Project Apollo, those who had made possible the day that had come. They clustered around displays and switchboards and dials and computers, monitoring every heartbeat of the spacecraft that had slipped out of the influence of its home planet and was now being enticed by the gravitational pull of the Moon. Most of them joined their friends and families in gathering around the televisions as well.

Among the black women watching the television, far from Mission Control, tucked away at a resort in the Poconos, Katherine Johnson divided her attention between the weekend leadership conference being held by her sorority, Alpha Kappa Alpha, and the fortunes of the Apollo 11 astronauts on their way to becoming the three most solitary beings in the history of humanity. As she watched the delicate dance of physics that propelled the Apollo capsule forward toward the Moon, her mind's eye superimposed equations and numbers upon each stage of the craft's journey, from launch to Earth orbit, from translunar injection to lunar orbit.

The intensity of the last few days at Langley had been matched only by the extreme heat that had enveloped the peninsula. It was nearly 96 degrees in Hampton that Saturday morning in July 1969 when Katherine and a car full of sorority members hit the road for the Poconos. It had been too hot to think, too hot to sleep, too hot to do anything except seek refuge anywhere you could find it, until the temperature ticked back down from intolerable to just bearable. The weekend escape had offered a break from both office and climate, each mile north taking her farther away from the steam heat that had held the area hostage for the last few days. Passing Washington, DC, she could breathe a little easier; by the time they crossed from Maryland into the foothills of Pennsylvania, the fever had broken, the air outside was crisper, the sky bluer and higher, the milder climate a reminder of her native West Virginia.

The Hillside Inn, perched on a grassy rise like an oversized farmhouse, was the perfect setting for the flock of pink-and-green-clad women who had convened for a weekend of planning and friendship. The sorority had tapped the most promising young women from collegiate chapters around the country so that they might learn from seasoned members like Katherine how best to organize the service projects that were at the core of their mission and activity. They talked about fundraisers for scholarships to black colleges, literacy campaigns, and voter registration drives. The kind of projects undertaken by the chapters around the country ranged from modest and one-off to sophisticated operations: one AKA chapter in Ohio ran a full-time job training center in one of the city's black communities.

The women doubled and tripled up in Hillside's thirty-three rooms, taking in the expanse of green and mountain views that were part of the region's iconic appeal. The inn's rustic luxury fulfilled the sorority's need for a quiet, reflective setting for their meeting. But it also boosted their racial pride: the Hillside Inn was the only resort in the Poconos with black owners. Albert Murray, a successful New York lawyer, had bought the land with his Jewish business partner in 1954. A year later the partner died, and Murray and his wife, Odetta, decided to use the property for a hotel. At the time, most resorts in the Poconos barred Negroes and even Jews, maintaining policies every bit as inflexible as the legal segregation of the South. The Hillside welcomed all guests, and most particularly wanted to provide upwardly mobile blacks with the same kind of vacation experience their white counterparts enjoyed.

The Hillside advertised in the *Norfolk Journal and Guide*, the *Pittsburgh Courier*, and *Ebony*; with its swimming pool and expansive 109-acre estate, it delivered on its promise of understated luxury. And of course, it distinguished itself with the things that the black sororities and social clubs and family reunions making the pilgrimages up Route 609 never would have found at the other retreats, even if they'd managed to get past the bellman—delights like the hearty southern-style home cooking. Three times a day, Katherine and her sorors sat next to each other family style in the inn's dining room, laughing and talking and debating over grits for breakfast, ribs and golden-fried chicken for lunch and dinner, and sweet potato pie and peach cobbler for dessert. The youngsters who staffed the dining room—all of them students at black colleges in the South, a conscious choice on the part of the Murrays—were constantly exposed to Hillside's professional class of patrons, examples in the flesh of what they might aspire to in their own lives.

Katherine loved the exacting standards of the women in the sorority; their shared desire to do things of value for other people, their fierce commitment to cultivate and display the best of the black community, served to deepen their personal bonds. They'd had to learn to work together to accomplish their goals, something that had served Katherine and the rest of the women well in their careers. The sorority had been a constant in her life since her days as a fifteen-year-old freshman at West

Virginia State; she had spent more weekends than she could remember attending sorority activities or meetings.

Katherine and the other women relaxed in each other's company in the intimate setting, enjoying it even more for the many years that the experience had been denied them. It was not yet so long ago since Katherine's father, Joshua, and Dorothy Vaughan's husband, Howard, had worked together, attending to the needs of the jet set at the Greenbrier, not so many summers past since Katherine herself had staffed the grand hotel's antiques store and served as a private maid to wealthy guests. It was just yesterday, it seemed, that *she* was the precocious adolescent learning to hold her own with the kitchen's French chef and making conversation with the president's brother and the other lofty guests who dropped in on the resort as part of their nomadic social circuit.

Those well-heeled people had all responded to something in the young bespectacled woman, something that gave them the feeling that she had a great future. Who among them would have ever imagined, however, that Katherine's future, and their country's future, and *the* future, as imagined by the likes of H. G. Wells and Jules Verne, would converge to be one and the same? Yet four days earlier, on July 16, 1969, fifty-year-old Katherine Johnson had been part of that group of insiders when the three-hundred-foot Saturn V rocket boosted the Apollo 11 craft and its three human occupants down the road to history.

Mission Control set the candle on fire at 9:37 a.m., early enough for the East Coast brain busters to take in the big event and get to work, then spend the rest of the day getting the color commentary. If the space shots hadn't exactly become commonplace since Alan Shepard's first foray, they happened often enough for talking heads like CBS's Walter Cronkite to wield the jargon of max Q and apogee and trans-Earth injection with the same nonchalance as the flight operations crew in the trenches of Mission Control. Still, the broadcasters knew—everyone in the audience knew—that even with twenty-six manned flights under NASA's belt, this was different, and they struggled to come up with superlatives to capture the moment. Cronkite gushed unabashedly, putting the magnitude into the context of the great machines of war and transportation that had transformed the American century: the mighty Saturn V rocket consumed the equivalent of ninety-eight railroad cars' worth of fuel; it propelled a craft that weighed as much as a nuclear submarine with the equivalent thrust of 543 fighter jets. The United States would spend $24 billion on Apollo, in order to plunge the sword into the heart of the Soviet Union's ambitions in space.

Not everyone shared Cronkite's exuberance. All that money—and for what? many wondered. So much money spent so that between 1969 and 1972 a dozen white men could take the express train to a lifeless world? Why, Negro women and men could barely go to the next *state* without worrying about predatory police, restaurants that refused to serve them, and service stations that wouldn't let them buy gas or use the bathroom. Now they wanted to talk about a white man on the Moon? "A rat done bit my sister Nell, with Whitey on the Moon," rapped performer Gil Scott-Heron in a song that stormed the airwaves that year.

At the beginning of the decade, the space program and the civil rights movement had shared a similar optimism, a certain idealism about American democracy and the country's newfound drive to distribute the blessings of democracy to all its citizens. On the cusp of the 1970s, as the space program approached its zenith, the civil rights movement—or rather many of the goals it had set out to achieve—were beginning to feel as if they were in a state of suspended animation. There were real and shining triumphs, certainly: the Civil Rights Act of 1964 and the Voting Rights Act of 1965 pried Jim Crow's legal grip off the country's workplaces, modes of transportation, public spaces, and voting box. But the economic and social mobility that had been held hostage by that legal discrimination remained stuck.

In the days leading up to the launch, two hundred protesters, led by the Reverend Ralph Abernathy, made their way to Cape Kennedy. Abernathy was Martin Luther King Jr.'s closest collaborator and had inherited the mantle of his Poor People's Campaign, the second phase of the civil rights movement. Abernathy and his fellow activists came to the launch site riding a mule train, challenging NASA's administrator, Tom Paine, on the worthiness of the space program when the poor and dispossessed in Watts and Detroit and rural Appalachia could barely put food on the table—assuming they had a house to put the table in. The Housing Rights Act of 1968, making it illegal to discriminate in the housing industry based on race, had lingered in Congress for years, vehemently opposed by legislators both in the North and the South. The bill only made it over the finish line in the wake of the 1968 assassination of Dr. King.

Katherine Johnson certainly knew all about the housing issue. Discrimination in housing remained the standard, but postwar economic mobility had given families like hers and Dorothy Vaughan's the means to move out of once-vibrant developments like Newsome Park and into comfortable, leafy, all-black subdivisions. The exit of professional families ruptured the connection that the less fortunate had to the world of college and middle-class jobs. Newsome Park and the hundreds of neighborhoods like it around the country became increasingly volatile, desperate islands where housing, schooling, and every other state-supplied service were left to deteriorate.

The decision to prioritize a victory in space over problems on Earth was the most widespread criticism against the space program. But even those voices in the black community who expressed admiration for the astronauts, who supported the program and its mission, took NASA to the woodshed for its lack of black faces. No black television commentators, no black administrators, no black faces in Mission Control, and most of all, no black astronauts. Blacks were still smarting over the perceived mistreatment of Ed Dwight, an astronaut trainee who was given his walking papers before he could even report for duty.

Though groups like ACD and Reentry Physics still employed several of the former West Computers, Katherine and others found themselves the only black employees in their branch. They were maybe less visible at work now that segregation had been ended. But they were perhaps more invisible professionally in the black community. The white NASA folks tended to live in enclaves, carpooling together and barbecuing together and sending their kids to school together. They talked about work and imported the hierarchies and nuances of their work lives into their neighborhoods.

The black NASA people spread out among other black professionals, where they were better known as the sorority sister or the member of the church choir or the diehard Hampton Institute alum who never missed a football game. Their neighbors might know they worked at NASA but have no concept of exactly what they did, or how close they were to the headline-grabbing events of the day. Because of the overwhelmingly white public face of the space program, the black engineers, scientists, and mathematicians who were deeply involved with the space race nevertheless lived in its shadows, even within the black community.

Katherine was sensitive to the disconnect. She, like Mary Jackson and many of Langley's other black employees, had worked hard for years to cultivate interest in math and science and space through the networks of their sororities and alumni associations and churches, with mixed results. In 1966, however, something had happened that looked like it might give them a tailwind.

Star Trek landed in American homes on September 8, 1966, an NBC network prime-time program. While NASA and the Project Gemini astronauts worked their way through twelve missions in the 1960s, in the fictional 2260s, the starship *Enterprise* set off from Earth on a peacekeeping and deep-space exploration mission, manned by a multinational, multiracial, mixed-gender crew. The corps, led by the suave, unflappable Captain James T. Kirk, included natives of an advanced United Earth, its history of poverty and war now in the past. Enemies in a former Earth Age labored side by side as colleagues and fellow citizens. Chekov, the Russian ensign; Sulu, the Japanese American helmsman; and the half-human, half-Vulcan first officer, Mr. Spock, added an interstellar touch of diversity. And there, on the bridge, a vision in a red minidress opened viewers' minds to what a truly democratic future might look like. Lieutenant Uhura, a black woman and proud citizen of the United States of Africa, served as the *Enterprise*'s communications officer.

Lieutenant Uhura, portrayed by the actress Nichelle Nichols, executed her duties with aplomb, managing the ship's communications with other ships and planets. When the first season ended in 1967, Nichols tendered her resignation to the show's creator, Gene Roddenberry, so that she could spend more time tending to her Broadway career. The producer, who wanted to keep Nichols in the cast, refused her resignation, asking her to take the weekend to mull it over.

That weekend Nichols attended a celebrity NAACP civil rights fundraiser in Los Angeles. One of the event's coordinators let her know that "her greatest fan," a fellow attendee, wanted to meet her. Expecting some eager, socially awkward adolescent, Nichols instead found herself face-to-face with Dr. Martin Luther King Jr.: King was a Trekkie! It was the only show that he and his wife, Coretta, allowed their children to watch, and he never missed an episode. Nichols thanked him for his effusive praise before mentioning almost casually that she had decided to leave the show. The words had barely escaped her lips before the Reverend interrupted her cold.

"You can't leave the show," King said to Nichols. "We are there because you are there." Black people have been imagined in the future, he continued, emphasizing to the actress how important and groundbreaking a fact that was. Furthermore, he told her, he had studied the Starfleet's command structure and believed that it mirrored that of the US Air Force, making Uhura—a black woman!—fourth in command of the ship.

"This is not a black role, this is not a female role," he said to her. "This is a unique role that brings to life what we are marching for: equality." The rest of Nichols' weekend was a fog of anger and sadness: what right did Dr. King have to upend her career plans? Eventually, she moved from resignation to conviction. Nichols returned to Gene Roddenberry's office on Monday morning and asked him to tear up the resignation letter.

How could Katherine not be a fan? Everything about space had fascinated her from the very beginning, and there, on television, was a black woman in space, doing her job and doing it well. A black person and a woman both, but also just Lieutenant Uhura, the most qualified person for the job. Katherine, in fact, thought science—and space—was the ideal place for talented people of any background. The results were what mattered, she told classrooms of students. Math was either right or wrong, and if you got it right, it didn't matter what color you were.

Star Trek was set in 2266, but it wasn't necessary to wait three centuries to see what America's finest minds could do, given free rein. The Apollo mission was happening now. In the Hillside Inn, among the group of her sorority sisters. Katherine gave in to the wonder of the moment, imagining herself in the astronauts' place. What emotions welled up from the depths of their hearts as they regarded their watery blue home from the void of space? How did it feel to be separated by a nearly unimaginable gulf from the rest of humanity yet carry the hopes, dreams, and fears of their entire species there with them in their tiny, vulnerable craft? Most people she knew wouldn't have traded places with the astronauts for all the gold in Fort Knox. The men existed all alone out there in the void of space, connected so tenuously to Earth, with the real possibility that something could go wrong. But given the chance to throw her lot in with the astronauts, Katherine Johnson would have packed her bags immediately. Even without the pressure of the space race, even without the mandate to beat the enemy. For Katherine Johnson, curiosity always bested fear.

The Eagle, the lunar lander, issued forth from the Apollo command module at 4:00 p.m. The touchdown caused a collective shiver. The crew was close, so close. The world waited for the door of the crab-like mechanical contraption to open. It took four hours. Then, finally, at 10:38 p.m.: sighs, applause, exuberance, dumbstruck silence, from all corners of the Earth, as Neil Armstrong planted his foot on luna firma. The actual landing had been the one part of the mission that had been impossible to rehearse prior to the actual moment—and the most dangerous. The Apollo 11 astronauts had given the mission only a middling chance of success: though Neil Armstrong handicapped the odds of returning to Earth safely at 90 percent, he thought they had only a 50-50 chance of landing on the Moon on the first go. Katherine Johnson had confidence: she knew her numbers were right, and she assumed that everyone else—Marge Hannah and the fellas there in her office, Mary

Jackson and Thomas Byrdsong and Jim Williams, everyone from the top of NASA to the bottom—had given their all to the mission.

Besides, Katherine always expected the best, even in the most difficult of situations. "You have to expect progress to be made," she told herself and anyone else who might ask. It had taken more than a decade of data sheets and plotting, IBM punch cards and long days and nights in front of the Friden calculator, delays and tragedies and most of all numbers; at this point, there were more numbers than even she could count. All on top of the long and monotonous years spent learning the basics of the machine that had given birth to the space program.

The trajectories of so many people had influenced hers along the way: Dorothy Vaughan and the women of West Computing. Virginia Tucker and all the women who had helped revolutionize aeronautics, with their work and their dogged presence at the NACA. Dr. Claytor and his enthusiastic preparation. John W. Davis from West Virginia State. Even A. Philip Randolph and Charles Hamilton Houston. Of course, it couldn't have happened without her parents. What she wouldn't have given for her father to see her—to see his baby girl who used to count the stars now sending men to travel among them. Joshua Coleman knew as if from second sight that Katherine, his brilliant, charismatic, inquisitive youngest child—a black girl from rural West Virginia, born at a time when the odds were more likely that she would die before age thirty-five than even finish high school—would somehow, someday, unite her story with the great epic of America.

And epic it was. Katherine allowed the moment, with all its implications, to sink in. There were still challenges ahead. She watched the men in the dust of the Moon and thought of the orbiting command service module, out of view of the camera, circling the Moon every ninety minutes. Neil Armstrong and Buzz Aldrin on the Moon would have a brief window to get back into the lunar lander and reconnect their dinghy with the mother ship above. After that, it would be three long days on the highway back to Earth, then through the fire of the atmosphere and into the terrestrial ocean below. Each leg carried the specter of the unknown; only after their landing matched the numbers of her equations, when they had been plucked from the ocean and cosseted in the waiting navy ship, would she be able to exhale.

But even then—only for a moment. Apollo still had six missions to go. And there was nothing like the thrill of the next thing. Katherine and Al Hamer had already started thinking about what it would take to plot a course to Mars; their colleagues Marge Hannah and John Young would look even farther into the cosmos, dreaming up a "grand tour" of the outer planets. It was built on the same idea as the rendezvous in the orbits of Earth and the Moon, where a spaceship doing a flyby of one planet used the planet's gravity to slingshot it on ahead to the next. The nimble minds in 1244 were already hopping from Mars to Jupiter and on to Saturn, like stones skipped in a glassy lake. One day, perhaps, the rest of humanity would follow them. Then, Katherine Johnson would *really* discover what was out there. It would be simple, she thought, just like sending a man into orbit around Earth, just like putting a man on the Moon. One thing built on the next. Katherine Johnson knew: once you took the first step, anything was possible.

EPILOGUE

It's the question that comes up most often when I tell people about the black women who worked as mathematicians at NASA: Why haven't I heard this story before? At this point, more than five years after I first began the research for what would become *Hidden Figures*, I've fielded the question more times than I can count. Most people are astonished that a history with such breadth and depth, involving so many women and linked directly to the twentieth century's defining moments, has flown below the radar for so long. There's something about this story that seems to resonate with people of all races, ethnicities, genders, ages, and backgrounds. It's a story of hope, that even among some of our country's harshest realities—legalized segregation, racial discrimination—there is evidence of the triumph of meritocracy, that each of us should be allowed to rise as far as our talent and hard work can take us.

The greatest encouragement along the way has come from black women. All too often their portrayals—our portrayals—in history are burdened with the negative imagery and vulnerability that come from being both black and female. More disheartening is how frequently we look into the national mirror to see no reflection at all, no discernible fingerprint on what is considered history with a capital *H*. For me, and I believe for many others, the story of the West Computers is so electrifying because it provides evidence of something that we've believed to be true, that we want with our entire beings to be true, but that we don't always know how to prove: that many numbers of black women have participated as protagonists in the epic of America.

Katherine Johnson's passion for her work was as strong during the remainder of her thirty-three-year career at Langley as it was the first day she was drafted into the Flight Research Division. "I loved every single day of it," she says. "There wasn't one day when I didn't wake up excited to go to work." She considers her work on the lunar rendezvous, prescribing the precise time at which the lunar lander needed to leave the Moon's surface in order to coincide and dock with the orbiting command service module, to be her greatest contribution to the space program. But another set of her calculations stood at the ready during the 1971 Apollo 13 crisis, when the electrical system of the spacecraft carrying astronauts Jim Lovell, Jack Swigert, and Fred Haise was crippled by an onboard explosion, making it impossible to run the guidance computer as programmed.

An astronaut stranded hundreds of thousands of miles from Earth is like a mariner from a previous age, adrift in the most remote part of the ocean. So what do you do when the computers go out? This was precisely the question Katherine and her colleague Al Hamer had asked in the late 1960s, during the most intense preparations for the first Moon landing. And in 1967, Johnson and Hamer coauthored the first of a series of reports describing a method for using visible stars to navigate a course without a guidance computer and ensure the space vehicle's safe return to Earth. This was the method that was available to the stranded astronauts aboard Apollo 13.

Before the crisis was over, however, even Katherine and Al's backup calculations would require a backup: from inside the spaceship, the glinting debris field from the damaged capsule was indistinguishable from the actual stars, making the method specified in Katherine's report impossible to use. Astronaut Jim Lovell used an even simpler calculation to tack his spaceship toward home, lining up the ship's optical sight with Earth's terminator, the line dividing the side of Earth that was in daylight from the shadow side, in nighttime. It was serendipitous that Lovell had taken the technique for a test run on Apollo 8 and knew how to make the calculations. What seemed like a routine check on a previous mission would save the crew's lives this time around. No one knew better than Katherine Johnson that luck favored the prepared.

Katherine Johnson worked with Al Hamer and John Young for the rest of her years at Langley, developing aspects of the space shuttle and the Earth resources satellite programs. But it is Katherine's connections to the most glorious and glamorous days of the space program that brought her to the public's attention. Every year since 1962, when John Glenn took to orbit, acclaim for Katherine Johnson's achievement grew. The black press—the *Norfolk Journal and Guide*, the *Pittsburgh Courier*, the *Amsterdam News, Jet Magazine*—embraced her even before John Glenn left Earth. Of NASA, *Amsterdam News* editor James Hicks wrote: "They are loud in their praise of a young West Virginia–born Negro girl who has prepared a science paper that was not only a key document in the flight of Commander Shepard into outer space but which will actually become 'THE' key document if and when we are able to put an astronaut into orbit." Over time, articles began to appear in the peninsula's *Daily Press* and in the *Richmond Times-Dispatch*, and Katherine's name became a necessary entry in any book detailing the accomplishments of black or female (or black female) scientists and engineers. Since the 1960s, she has been invited into classrooms to inspire students with the stories of how mathematics has defined her life. In recent years she has become too fragile to make the trips to visit students; on August 26, 2016, she will be ninety-eight years old. Now the students come to her, making pilgrimages to see her in the retirement residence where she lives. Her contributions to the space program's signature epoch earned her NASA Group Achievement Awards for Project Apollo and the Lunar Orbiter Project. She has received three honorary doctorates and a citation from the state of Virginia. And a charter high school in North

Carolina has launched a STEM institute bearing her name. In 2015, President Obama awarded Katherine Johnson the Presidential Medal of Freedom, an honor that astronaut John Glenn received in 2012.

Katherine Johnson is the most recognized of all the NASA human computers, black or white. The power of her story is such that many accounts incorrectly credit her with being the first black woman to work as a mathematician at NASA, or the only black woman to have held the job. She is often mistakenly reported as having been sent to the "all-male" Flight Research Division, a group that included four other female mathematicians, one of whom was also black. One account implied that her calculations singlehandedly saved the Apollo 13 mission.

That even Katherine Johnson's remarkable achievements can't quite match some of the myths that have grown up around her is a sign of the strength of the vacuum caused by the long absence of African Americans from mainstream history. For too long, history has imposed a binary condition on its black citizens: either nameless or renowned, menial or exceptional, passive recipients of the forces of history or superheroes who acquire mythic status not just because of their deeds but because of their scarcity. The power of the history of NASA's black computers is that even the Firsts weren't the Onlies.

No one is more in agreement with this point of view than Katherine Johnson. It was from her descriptions to me in interviews of the West Computing office that I first had an inkling of just how many black women might have worked at Langley. I first heard Dorothy Vaughan's name from Katherine, and no one—not even the brainy fellas—merited more of Katherine's admiration than Dot Vaughan. Of Margery Hannah, West Computing's first supervisor, who eventually joined Katherine's branch, she said, "She was extremely smart, and she didn't get half the credit she deserved." She enjoyed bragging about Christine Darden's accomplishments more than she wanted to talk about her own work. "I never go into a school without mentioning Christine," she told me. She is generous in her appreciation of other people's talents in the way of someone who is in full command of her own gifts. As much as Katherine Johnson's technical brilliance, it's her personal story and her character that shine on us like a beacon. What could be more American than the story of a gifted little girl who counted her way from White Sulphur Springs, West Virginia, to the stars? That along the way she equaled the prowess of an electronic computer, becoming a brainy, female John Henry, only served to burnish her myth. She is charismatic and self-possessed, cool under pressure, independent-minded, charming, and gracious. Her unencumbered embrace of equality, applying it to herself without insecurity and to others with the full expectation of reciprocity, is a reflection of the America we want to be. She has been standing in the future for years, waiting for the rest of us to catch up.

But perhaps most important, Katherine Johnson's story can be a doorway to the stories of all the other women, black and white, whose contributions have been overlooked. By recognizing the full complement of extraordinary ordinary women who have contributed to the success of NASA, we can change our understanding of their abilities from the exception to the rule. Their goal wasn't to stand out because of their differences; it was to fit in because of their talent. Like the men they worked for, and the men they sent hurtling off into the atmosphere, they were just doing their jobs. I think Katherine would appreciate that.

For Mary Jackson, who remained steadfast in her pursuit of the ideals of the Double V—for African Americans and for women—the years following the Moon landing would be a time of change and choice. "Rockets, moon shots, spend it on the have-nots," Marvin Gaye sang in his 1971 anthem "Inner City Blues," addressing the quagmire that was Vietnam, an economy beset by inflation, and most of all, the isolation, anger, and economic despair of blacks living in Detroit, Washington, DC, Watts, and Baltimore. In the 1960s, it had felt possible that the idealism of Camelot, the Great Society, and the civil rights movement, those inheritors of the Double V, might finally wash away the poverty and injustice that had plagued America since its founding. As the decade drew to a close, it became apparent that the dream of Dr. King that had rung out over the Lincoln Memorial was actually the explosive dream deferred of Langston Hughes' poem "Harlem." "What happens to a dream deferred? Does it dry up like a raisin in the sun?" . . . In

Newsome Park, there was dwindling evidence of the hopefulness that Eric Epps had displayed when he dedicated the development's community center in 1945. The spaceflight revolution had solidified Katherine Johnson's and Dorothy Vaughan's positions in the middle class, but the neighborhood they and Eunice Smith and many others left behind was more and more like a poor island, cut off from the jobs and schools that would help them make the same leap the West Computers had made.

And that was before getting to "pollution, ecological damage, energy shortages, and the arms race," the gremlins of the century's technological revolution. Instead of creating unifying hope, an expansive space program was "salt on the wounds of the country's more Earthbound concerns," wrote NASA historian Robert Ferguson. As early as 1966, President Johnson, the space program's biggest political champion, began looking at NASA as a "big fat money pot" that he could drain to ease a budget strapped by social programs and Vietnam. With the Moon landing achieved, the victory over the Soviet Union in hand, there was no urgency to push beyond Project Apollo, whose last two missions narrowly escaped cancellation.

The press surrounding the end of the Apollo program was clamorous, but the cancellation of another program also garnered headlines. In 1972, the United States decided to cancel its supersonic transport program, the SST, which many aerodynamicists had hoped would give them an "Apollo moment," a glorious, high-profile display of their technology. The expensive program raised the hackles of those concerned about its negative impact on the Earth's ozone layer, but it was the sonic boom "carpet" that swept across the landscape as the plane passed overhead that really inflamed public opinion. Reports claimed that shock waves from the high-speed commercial planes were "frightening residents, breaking windows, cracking plaster, and setting dogs to barking." Some purported that the invisible menace had even caused the "death of pets and the insanity of livestock." Local authorities received complaints of broken windows and traumatized animals, and calls to police surged as citizens reported unidentified blasts that came literally out of the clear blue sky.

The supersonic and hypersonic transport machines dreamed up in the 1950s and 1960s would have to wait, although in the 1970s Langley did turn much of its focus back to NASA's first *A*: aeronautics. "In 1969 alone, there were 57 certified American airlines, which carried approximately 164 million originating passengers and some 20 billion revenue ton-miles of freight," NASA revealed in a 1971 publication. The aerodynamicists' priorities for the new decade were less glamorous, but a necessary part of solving the problems that were the result of an increasingly mobile society. One of the problems that the center focused on was noise abatement: busy skies were often noisy skies, even without sonic booms. Another issue was efficiency. With increasing fuel prices, the aircraft industry shifted its priority from increasing speed and power to boosting efficiency in subsonic or low supersonic flight.

Langley announced a sweeping reorganization in 1970, decreasing its workforce to a total of 3,853 from its peak of 4,485 employees in 1965. For those who lived through the reorganization, announced in the form of a forty-seven-page avocado-green book that landed on employees' desks at the end of September that year, it was in many ways a more jarring time than the period of transition from the NACA to NASA. Waves of RIFs and RIGs—Reductions in Force and Reductions in Grade—happened so frequently at Langley in the 1970s that they spawned a new verb, as in "John got riffed last week." Those who did survive the RIFs felt a sense of betrayal at NASA's significantly reduced ambitions. Not only were the brain busters not heading to Mars and the outer planets, but by December 1972, they had left their final footprints on the Moon. The summit of humanity's knowledge crashed into low-orbit reality. The NASA of the 1970s was interested in "routine, quick-reaction and economical access to space." The agency would never return to the glory of the Apollo years. But despite the downsizing of everything—budgets, workforce, expectations—the will to explore the world beyond Earth's atmosphere did not, would not, could not go away.

Mary Jackson managed to surf Langley's turmoil even as the sections, branches, and divisions around her recombined with greater frequency, the work groups at the bottom of the organization chart transforming like shards in a great NASA kaleidoscope. The names changed—Compressibility, Aero-Thermo, Applied Theory, Large Supersonic Tunnels, Transonic Aerodynamics, High-Speed Aircraft, Subsonic-Transonic—but her partnership with Kazimierz Czarnecki remained a constant. She stayed focused on the research she had pursued since becoming an engineer in 1958: the investigation of the impact of roughness (such as rivets or grooves) on the surface of a moving object on the boundary layer, that thin layer of air that passes most closely over a moving object. Never one to miss an opportunity to continue her education, Mary took FORTRAN classes, teaching herself to program. The computers that had made long-distance spaceflight possible were also revolutionizing aeronautical research, a specialty known as computational fluid dynamics. The engineers now conducted experiments in their beloved wind tunnels and then compared the results with simulations on their computers. Just as the electronic machines had taken the place of human computers in aeronautical research, the day would eventually come when the computer would displace the wind tunnel itself.

Mary Jackson was a tireless promoter of science and engineering as a meaningful and stable career choice. She made so many speeches at local schools that one might have thought she was running for office: Thorpe and Sprately Junior High Schools, Carver and Huntington High Schools, Hampton Institute, Virginia Wesleyan, a small college in Norfolk. At the King Street Community Center, where Mary had worked as the USO secretary during World War II, she started an after-school science club for junior and senior high school students. She helped the students build a smoke tunnel and conduct experiments, and taught them how to use the tool they created to observe the airflow over a variety of airfoils. "We have to do something like this to get them interested in science," Mary commented in a 1976 article in the employee newsletter *Langley Researcher*, which profiled her for being honored as the center's Volunteer of the Year. "Many times, when children enter school they shun mathematics and science during the years when they should be learning the basics."

In 1979, Mary Jackson organized the retirement party for Kazimierz Czarnecki, who was leaving government service after forty years.

Two years prior, the facility that had been the bedrock of most of their work—the Four-by-Four-Foot Supersonic Pressure Tunnel, the third member of Mary and Kaz's partnership—had come to the end of its service at Langley as well. In 1977, the tunnel that had been state-of-the-art technology when it began operations in 1947 was razed to make way for the National Transonic Facility, a 1.2 Mach, $85 million tunnel that was powered by cryogenic nitrogen.

It was a moment for Mary to reflect on her career. She traveled regularly to make presentations at industry conferences, and by the end of the 1970s she had twelve authored or coauthored papers to her name. She had progressed from computer to mathematician to engineer, and in 1968 had been promoted to the level of GS-12. The budget cuts and RIFs of the 1970s made promotions harder to come by, however, and the next rung on the ladder for Mary Jackson—GS-13—was starting to look distant. GS-13 was a significant threshold, with few women in Langley at that grade in the mid-1970s. This was a contrast with Goddard, where both Dorothy Hoover and Melba Roy had hit the GS-13 mark by 1962. In 1972, NASA's agencywide goal was "to place a woman in at least one of out of every five vacancies filled at levels GS-13 through GS-15." The numbers of women, professional and administrative, had grown along with Langley's general level of employment, but women were still a scarcity in high-level technical positions and in management. Even seemingly small barriers conspired to keep larger numbers of women from advancing: until 1967, the Langley Field golf course—as in other workplaces, a prime location for networking—restricted women to playing during the workday, rather than allowing them to golf alongside men after work.

In 1979, Mary Jackson was fifty-eight years old and coming to the conclusion that she had probably hit the glass ceiling. It would have been easy for her to reap the benefits of seniority, reducing her workload and taking a long coast toward retirement. Even if the next promotion eluded her, she still had the prestige of being an engineer and the satisfaction of knowing how hard she had worked to arrive at this point. But a position opened up in the Human Resources Division, and Mary's name was floated to fill it: Federal Women's Program Manager, charged with pushing for the advancement of all of the women at the center. To relinquish her hard-won title of engineer, at an organization that was created and run by engineers, was no easy choice.

Mary's career frustration wasn't unique, she knew. When she looked around, she saw many women and minorities at Langley trapped in the sticky middle grades, unable to rise to the level that their ability would otherwise merit. Did Langley really need one more GS-12 aeronautical engineer, even if the seat was occupied by a black woman? Or would the center be better served by someone who could help make way for legions of employees, at every level and from every background, liberated to give their best to their work? Mary Jackson wasn't wired to take the easy road or be satisfied with the status quo. If the decision wasn't simple, it was certainly clear. Stepping off the engineering track wasn't a sacrifice if it allowed her to act on her principles. Taking a demotion from GS-12 to GS-11 in order to accept the less-prestigious position, Mary Jackson threw herself in 1979 into her new role as the center's Federal Women's Program Manager.

Helping girls and women advance was at the core of Mary's humanitarian spirit; she saw the relationships between women as a natural way to bridge racial differences. She had been instrumental in bringing the separate regional white and black Girl Scout Councils together into a unified service organization for all girls in southeastern Virginia. In 1972, Mary volunteered as an equal opportunity employment counselor, and in 1973 she joined Langley's Federal Women's Program Advisory Committee. Both programs had been created in the 1960s to make sure that the federal government was hiring and promoting without differentiation by race, gender, or national identity. At Langley, as in other federal workplaces, the programs had a beneficial secondary effect: they gave female and minority employees a formalized way to make connections and boosted their centerwide visibility. Mary had always been a natural networker, bringing people together to help one another and to marshal their support for the many causes that were dear to her heart. She became an energetic member of a group of Langley women who were determined to push for opportunities for women of all colors at NASA, clearing the way for women to take their place as equals alongside men in science and engineering jobs, and also looking for ways to help secretaries and clerical employees to make the leap into technical jobs and program management. Accepting the position as the Federal Women's Program Manager was a way of uniting twenty-eight years of work at Langley with a lifetime commitment to equality for all.

One of the most difficult aspects of writing a book is knowing that there's not enough space or time to give voice to all the incredible people you meet along the way. The original draft of *Hidden Figures* had a final section portraying in detail how Mary Jackson and her fellow travelers went to all lengths in the 1970s and 1980s to extinguish the lingering traces of what NASA historian Sylvia Fries called the "fantasy that men were uniquely gifted to be engineers." Like Mary, the final narrative stepped away from the daily routines of research to follow the women of Langley as they formed alliances and used all the ingenuity they brought to engineering to change the face of the center's workforce. Making the decision to trim this section was difficult; while it allowed for the chance to spend more time with Dorothy, Mary, and Katherine in the golden age of aeronautics and space, it meant ending the book before Mary's decision to leave engineering for Human Resources. It also meant saying good-bye to one of my favorite "characters" in this sweeping drama, who has become a treasured friend in real life: Gloria Champine. The relationship between Gloria and Mary, which grew out of Mary's decision to sacrifice her engineering career for the future career prospects of other women, is one of the most poignant of all the stories I uncovered in the research.

Gloria Champine was born at Fort Monroe in Hampton in 1932, her family home a stone's throw from Mary's. Her father was an airman at Langley Field who was instrumental in the development of the parachute. He died in the crash of a Keystone bomber on a flight from Langley in 1933. Her stepfather was the crew chief on the only XB-15 ever built, which was stationed at Langley. Gloria spent part of her childhood on the base, where "everybody's daddy had a plane." She grew up overhearing her stepfather and his crew tell stories of the "crazy things" the NACA nuts put them through in order to analyze the flying qualities of their experimental model bomber. Gloria, who is white, graduated from Hampton High School in 1947, completed an associate's degree at a local business college, and found a job as the secretary to the head of a printing company in Newport News. In 1959, Gloria took the civil service exam and accepted a job as a secretary in the Mercury range office, helping with the logistics required to build the worldwide tracking network that debuted with John Glenn's orbital flight.

In 1974, an equal opportunity program gave Gloria the chance to advance from a clerical position in the Dynamic Loads Division into a faster-track administrative position in the Acoustics Division. Then, she competed for an even higher position as the Technical Assistant to the Division Chief of Space Systems, a job that had previously only been held by men. She went through the interview process three times, and each time she came out on top. "They kept testing you because they didn't want to give the position to a woman," a friend in Human Resources confided to her. Eventually, however, the center was obligated to hire Gloria: the best candidate for the job, the first woman in the position.

When Mary and Gloria were girls in the early part of the twentieth century, only the most gifted seer could have predicted the changes that would bring their paths together. In later years, Mary would describe to Gloria the segregation she had experienced in the early years at Langley. They met through one of the Federal Women's Program committees and became friends, collaborators, and conspirators in the service of a shared belief in helping unrecognized talent get its day in the sun. Like Mary, Gloria Champine had a "hard head and strong shoulders and back." She couldn't keep herself from acting when she saw a way to give someone else a leg up. She always kept an extra women's blazer behind the door in her office, in case a potential job candidate needed a little sartorial sharpening to make a better impression. When a young black woman who spent the summer interning with her mentioned an interest in computers, Gloria marched her over to meet the head of a programming branch in the Business Systems Division. The young woman secured a place in a programmer trainee program.

Male supervisors warned Gloria to "stay away from the woman stuff," but the woman stuff was just as important to Gloria as it was to Mary Jackson. She had seen how dependent her mother, who was smart but valued for her beauty, had been on her father and stepfather. Gloria vowed never to be in the same situation; she never entertained the idea of not working, even after her three children came along. It was a decision that helped her to bear up when she separated from and then divorced her husband in the mid-1960s, leaving her a single mother and the head of her household at a time when the majority of white women still didn't work outside the home.

In 1981, Langley sent Mary Jackson to NASA headquarters in Washington, DC, for a year of training to become an equal opportunity specialist. Mary had already decided who should follow her as Langley's Federal Women's Program Manager. Though Gloria didn't come from a technical background, her military upbringing and fifteen years of experience at NASA gave her an understanding of the business of engineering and the motivations of the engineers. She knew airplanes better than a lot of the engineers she worked with. She was also a quick study with computers: Mary Jackson taught her how to "reprogram" the computers in the Human Resources Division, going deep into the databases that fed the systems in order to run statistical reports on employee qualifications and promotions. These reports revealed that female graduates with the same degrees as men were still more often hired as "data analysts," the upgraded term for the center's mathematicians, than as engineers. Black employees with similar qualifications lagged their white counterparts in promotions and were more likely to be steered to support roles, such as work in the Analysis and Computation Division, where Dorothy Vaughan had been reassigned, than to engineering groups. She showed Gloria how lacking a single course on a college transcript, such as Differential Equations, could keep an otherwise qualified and well-reviewed woman from keeping up with her male counterparts, even years after she had entered the workforce.

For the next five years, Mary Jackson and Gloria Champine were an effective social engineering team within the Equal Opportunity and Federal Women's Program offices. For three of those five years, they worked for my father, Robert Benjamin Lee III, a research scientist in Langley's Atmospheric Sciences Division. My father's move into equal opportunity was part of a career development program designed to "season" him for moves into management when he returned to his division.

Mary, however, spent the rest of her career in the equal opportunity office, retiring in 1985. Her husband, Levi Jackson Sr., had spent the end of his working years at Langley as well, transferring from the air force base in the 1980s, still working as a painter. "We always thought it was so cool that Grandma worked in the wind tunnels and Grand-daddy painted them," remembered their granddaughter, Wanda Jackson. To the end of his life, Levi Jackson was devoted to Mary and proud of her every achievement. Mary stayed as busy over the next twenty years as she had been over the previous sixty-four, filling her days with her grandchildren and the volunteer work that gave her so much ful-fillment. Mary Jackson died in 2005, and Gloria Champine penned a moving obituary that was published on the NASA website. "The penin-sula recently lost a woman of courage, a most gracious heroine, Mary Winston Jackson," Gloria wrote. "She was a role model of the highest character, and through her quiet, behind-the-scenes efforts managed to help many minorities and women reach their highest potential through promotions and movement into supervisory positions."

Gloria, too, ended her thirty-year career at Langley in the equal op-portunity office, building on Mary's legacy, making sure that no talent at Langley was overlooked. One of those whose careers she tracked was Christine Darden, the young mathematician who had been galvanized by Sputnik back in 1957. Christine's first years at Langley had been an exercise in enduring monotony. Though the Reentry Physics Branch had been an exciting place in the run-up to Apollo, long development lead times meant that by the time Christine came to the office, most of the interesting work had been completed, and the pace had slowed significantly. Although Christine's pool was attached to an engineering group, most days she felt that she had entered a time machine. Many of the women in the pool of data analysts were former West Computers, and even though Christine had significant FORTRAN programming experience from her time in graduate school, a Friden calculator sat on her desk awaiting her input, just as it had for the computers in the 1940s. It was "deadly," she said. She was working for the organization that had just led the charge to the Moon, and yet in her corner of NASA, Christine felt like the future had passed her by.

It took persistence, luck, and more than a little cheek to break out of what had become such tedium that Christine thought many times about quitting. She had survived the Green Book RIF in 1970, but just prior to a second wave in 1972, she happened to overhear her boss talking to someone in the Human Resources Department: she was on the hit list! In the complex game of RIF chess, she was being knocked off the board by a black man who had been hired at the same time as her, but as a mathematician. He had been sent off to an engineering group and promoted; she, with less seniority, was slated for layoff.

The revelation spurred her to action. Rather than raise the issue with her boss, Christine went directly to the division chief—her boss's boss's boss, none other than John Becker, Langley's éminence grise, now on the cusp of retirement.

"Why is it that men get placed into engineering groups while women are sent to the computing pools?" Christine asked him. "Well, nobody's ever complained," he answered. "The women seem to be happy doing that, so that's just what they do." Becker was a man from another era. His wife, Rowena Becker, had been an "excellent mathematician"—the two met during the war, in the eight-foot tunnel—but after marrying she made the decision to leave Langley to become a full-time wife and mother. His frame of reference for working women and their expectations was like that of most men of his generation. But just as Becker had been willing to admit he was wrong when challenged by Mary Jackson in the 1950s, he rose to meet Christine Darden's challenge twenty years later. Two weeks after Christine walked into John Becker's office, she was assigned to a group working on sonic boom research.

Christine's new boss, David Fetterman, was a self-described "wing man" who had decided to stay in aeronautics even as others were moving to space. He was happy working on his research independently and assumed that his new charge felt the same way. So he handed Christine a fly-or-fail research assignment: she was to take the industry standard algorithm used to minimize sonic boom for a given airplane configuration (developed by Cornell researchers Richard Seabass and Albert George) and write a FORTRAN program based on it. It was work at aeronautics' leading edge, a computational fluid dynamics project that might help to mitigate the sonic boom that had made commercial supersonic flight so unpalatable.

It took three years of work, but the results were published in a 1975 paper entitled "Minimization of Sonic-Boom Parameters in Real and Isothermal Atmospheres." Christine was the sole author. The code she wrote as an aspiring engineer is still the core of sonic boom minimization programs that aerodynamicists use today. It was an important contribution and a career-making achievement, but the road from that breakthrough moment to becoming an internationally recognized sonic boom expert with sixty technical publications and presentations under her belt and a member of NASA's Senior Executive Service was still not direct.

In 1973, Christine took a computer programming course through Langley's partnership with George Washington University. She had excelled at Hampton Institute, powered through her master's degree at Virginia State, and finally landed a position with an engineering group at NASA, but the class of eight students—seven white and one black, seven men and one woman—was the first time she had been in an integrated school setting. She was intimidated at first, but high marks in the class made her decide to pursue a doctorate. Getting approval to enroll in the program took some doing. An upper-level supervisor denied her initial request. Even after she got the approval, she still was "juggling the duties of Girl Scout mom, Sunday school teacher, trips to music lessons, and homemaker," for her two daughters in addition to her full-time work at Langley.

The doctorate in mechanical engineering took ten years. It was bestowed upon her in 1983, forty years after the first West Computers walked into Langley. Christine's success was supported by the work of the women who had come before her, her research based on the uncountable number of numbers that had passed through their hands and minds. Even with two of the finest credentials in the field to her name—a PhD and a major research contribution—it would take one more push before Langley acknowledged Christine Darden with the promotion that matched her accomplishments.

Gloria Champine admired Christine Darden's intelligence and the dogged way she had pursued her PhD. From her perch in the equal opportunity office, she knew that women at the center—even at the top levels—were still being leapfrogged by men, and Christine was one of them. By the mid-1980s, Christine had moved up to GS-13, but even with the doctorate, she was having problems breaking into GS-14. On the other hand, a white male engineer who had started at the same time, with similar quality performance reviews, had already hit the GS-15 level. Gloria knew the Langley way: "Present your case, build it, sell it so they believe it." She created a bar chart and showed it to the head of her directorate—a manager one level down from the top of Langley—who was shocked at the disparity. With Gloria's efforts, the promotion came, and after it, the renown and the mobility for Christine that should come to people with such outstanding abilities. It was one of Gloria's proudest moments. Christine had already done the work; Langley just needed someone who could help it see the hidden figures.

"What I changed, I could; what I couldn't, I endured," Dorothy Vaughan told historian Beverly Golemba in 1992. Dorothy retired in 1971 after twenty-eight years of service. The world had changed dramatically since the day she had taken the bus from Farmville to the war boomtown, but not quite enough to fulfill her last career ambitions. The Green Book landed on Dorothy's desk just two days after her sixtieth birthday. Her name was in the book, but it was not where she hoped it would be.

"It involved a promotion," Dorothy's daughter, Ann Vaughan Hammond, told me, though to exactly what, her mother never said. Dorothy played it close to the vest, giving her family only the barest sketch of her final disappointment. In all likelihood, she had expected to serve out her last few years as a section head, regaining the title she had held from 1951 through 1958. What a triumph it would have been to return to management, but as the head of a section that employed both men and women, black and white. The section head position was given to Roger Butler, a white man, who also held the post of branch chief. Sara Bullock, the East Computer who had been put in charge of the group programming the Bell computer back in 1947, was appointed head of one of the branch's four sections. Bullock was one of the rare female supervisors, particularly outside of administration. In 1971, there were still no female branch chiefs, no female division chiefs, no female directors at Langley.

And for the first time in almost three decades, no Dorothy Vaughan. Dorothy Vaughan's time as a supervisor back in the 1950s was relatively brief, but during those years she had midwifed many careers. Her name never appeared on a single research report, but she had contributed, directly or indirectly, to scores of them. Only reluctantly did she agree to a retirement party; she never liked it when people made a fuss. She discouraged her family from coming and found a ride (despite all her years in automobile-dependent Virginia, she never did learn to drive). Many of her new colleagues turned out to celebrate her, including her boss, Roger Butler. Of course, many of her old colleagues came too. Once upon a time they were girls who had come to Langley expecting a six-month war job; now they were older women with decades of membership in an elite scientific club. At one point in the evening, Lessie Hunter, Willianna Smith, and other West Computers gathered around their former supervisor for a picture, which was published the next week in the *Langley Researcher*. It was perhaps the only photographic evidence of the story that began in May 1943 with the Band of Sisters in the Warehouse Building. Though Langley was meticulous about documenting its employees over the years, individually and in groups, I have yet to stumble upon another Langley photo of the West Computing section.

Dorothy Vaughan had always loved to travel, and in retirement she indulged herself, traveling for pleasure across the United States and to Europe. In her eighties she took a trip to Amsterdam with her family. At home, she remained as frugal as she had been during the Depression and the war, never spending when she could save, never discarding what she could salvage.

At some point, some years into her retirement, a woman came to the house, trying to enlist her in a class-action lawsuit over pay discrimination against the women who had worked at Langley. Dorothy sat on her couch and gave the woman a polite hearing, and then said: "They paid me what they said they were going to pay me," and that was the end of that. She never had been one to dwell on the past. After her retirement party, Dorothy Vaughan never went back to Langley. The photo album, the service awards, and the retirement gifts—all of them she tucked away in the keepsakes box in the back of the closet. The greatest part of her legacy—Christine Darden and the generation of younger women who were standing on the shoulders of the West Computers—was still in the office.

ACKNOWLEDGMENTS

The title of this book is something of a misnomer. The history that has come together in these pages wasn't so much hidden as unseen—fragments patiently biding their time in footnotes and family anecdotes and musty folders before returning to view. My first thanks are to the historians and archivists who helped me reconstruct this story through its documents: to Colin Fries at the NASA History Office in Washington, DC, Patrick Connelly at the National Archives and Records Administration (NARA) Philadelphia, Meg Hacker at NARA Fort Worth, Kimberly Gentile at the National Personnel Records Center, and Tab Lewis at NARA College Park. Thanks also to Donzella Maupin and Andreese Scott at the Hampton University Archives, and Ellen Hassig Ressmeyer and Janice Young at West Virginia State University's Drain-Jordan Library.

I've been buoyed by the enthusiasm of David Bearinger and Jeanne Siler at the Virginia Foundation for the Humanities since the day I walked into their office unannounced, in the middle of an early spring blizzard in 2014. Because of their support, the Human Computer Project, which sprang out of my research for the book, will be able to pick up the baton from *Hidden Figures* by creating a comprehensive database of all the female mathematicians who worked at the NACA and NASA during the agency's golden age. Thanks to Doron Weber at the Sloan Foundation, who was willing to take a flyer on a first-time author; Sloan's support made it possible for me to make recovering this important history a full-time job.

I couldn't have had a better team to work with at William Morrow. Trish Daly, even though you are off to new ventures, I'll always be grateful for the dogged way you worked to put *Hidden Figures* at the top of your list. Rachel Kahan, thank you for your calm guidance in helping me bring this book home. To have a book published is exciting enough; to have it made into a film at the same time is truly a once-in-a-lifetime opportunity. Thanks to my film agent, Jason Richman, at the United Talent Agency, my lawyer, Kirk Schroeder, and especially to Donna Gigliotti, *Hidden Figures'* producer, who was able to see a movie in a fifty-five-page book proposal. She's one of the most gifted professionals I've ever met, in any industry.

Nowhere has *Hidden Figures* received a warmer reception than in my hometown, Hampton, Virginia. My deepest gratitude to Audrey Williams, president of the Hampton Roads Chapter of the Association for the Study of African American Life and History (ASALH), which served as sponsor for the seed stage of the Human Computer Project. Thank you to Mike Cobb and Luci Coltrane of the Hampton History Museum for inviting me to be part of the museum's speaker series, and to Wythe Holt and Chauncey Brown for their vivid recollections of the early days of life in Hampton, which added wonderful detail and texture to the book's narrative.

Current and former employees of the Langley Research Center, people too numerous to mention in such a limited space, have supported this project in many ways over the last few years, including Gail Langevin, NASA Langley's History Liaison. Andrea Bynum invited me to present my research in progress at Langley's Women's History Month celebration in March 2014 and has been a tireless supporter of the book ever since. Mary Gainer Hurst, Langley's recently retired Historic Preservation Officer, is a heroic public historian; thanks to her, thousands of interviews, wind tunnel test logs, photos, personnel documents, org charts, articles, and other primary materials that bear witness to Langley's extraordinary history are available to the public via the NASA Langley Cultural Resources website and related YouTube channel. So much of the connective tissue of this story came from the untold hours I spent consulting the information she so expertly recovered and curated.

Belinda Adams, Jane Hess, Janet Mackenzie, Sharon Stack, and Donna Speller Turner all shared recollections both of the technical work they were involved in and of the changing opportunities for women at Langley over the years. Harold Beck and Jerry Woodfill entertained my technical questions regarding the months leading up to John Glenn's orbital flight and the crisis of Apollo 13, respectively. My interview with engineer Thomas Byrdsong, who reminisced about being one of Langley's first black male engineers, is a bittersweet memory because it occurred less than a month before he passed away.

This book would not have been possible without the cooperation and support of the women who lived the history, and their friends, families, and colleagues. Bonnie Kathaleen Land, my former Sunday School teacher, has the distinction of being the very first person I interviewed for this book, in 2010; she passed away in 2012 at ninety-six years old. Thanks to Ellen Strother, Wanda Jackson, and Janice "Jay" Johnson for the marvelous tales of Mary Jackson's rich and active life outside the office.

Though Gloria Rhodes Champine's story appears only in the epilogue, there are many chapters in the book that bear her fingerprints. Her understanding of Langley's airplanes, its culture, and its people have been indispensable to helping me tell this story. Christine Darden is at once enormously talented and disarmingly modest, and it's a great source of pride that I've learned enough about aerodynamics over the course of this research to appreciate the scale of her achievements. Thanks to both of them for the wisdom and encouragement they have given me since *Hidden Figures'* beginning.

Ann Vaughan Hammond, Leonard Vaughan, and Kenneth Vaughan were instrumental in helping me reconstruct the details of their mother, Dorothy Vaughan's, early life and the trajectory that brought her to Langley. I thank them for allowing me to get to know her through their eyes.

Jim Johnson and his stories of serving in the Korean War were firsthand evidence of the enduring power of the Double V. Joylette Goble Hylick and Katherine Goble Moore have my utmost admiration for all they have done to preserve the legacy of their mother and the other women whose talents formed the basis for the most rewarding work I've ever done.

The lessons I've learned from Katherine Coleman Goble Johnson could fill another book. Her generosity in sharing her life story with me has changed my life, and for that I will be forever grateful.

Most of this book was written in Valle de Bravo, Mexico; my thanks to all the friends who offered support and encouragement each day. I'm indebted to my "kitchen cabinet," the people who over the course of the last six years have helped me carry this project over the finish line. My thanks to all of the friends there who offered daily encouragement and support, especially Marcela Diaz, Jim Duncan, Larry Peterson and Sabine Persicke. Particular mention must go to Margot Lopez who generously lent me her studio whenever I needed a quiet place to meet a deadline. Melanie Adams, Jeffrey Harris, Regina Oliver, Chadra Pittman, and Danielle Wynn have been my hometown cheering squad, never too busy to share contacts, suggestions, or a sympathetic ear. Susan Hand Shetterly, Robert Shetterly, Gail Page, and Caitlin Shetterly never failed to provide insights, wonderful meals, and quiet writing nooks. My siblings Ben Lee, Lauren Lee Colley, and Jocelyn Lee have been a constant source of inspiration, memories, and encouragement.

From our very first conversation, my literary agent, Mackenzie Brady Watson, has been one of *Hidden Figures*' greatest champions. Her expansive vision and business instincts have helped give this story a platform beyond anything I could have imagined.

As the child of a Hampton University English professor and a NASA research scientist, it was probably inevitable that I would eventually write a book about scientists. Drs. Margaret G. Lee and Robert B. Lee III have made telephone calls on my behalf, set up interviews, arranged meetings, scoured their memories for names and events, offered context and suggestions for telling the history, attended my presentations, made early morning and late-night runs to the airport, received packages, graciously allowed me to turn their home into an office, and supported my writing in countless other ways. Mommy and Daddy, I love you more than words can ever say.

Finally, no one has given more to this project than my husband, Aran Shetterly. He has read every version of *Hidden Figures* starting with the very first draft of the book proposal, improving it at every step along the way with his fierce intelligence and editorial savvy. His experience as a writer and researcher has been invaluable in terms of helping me figure out how to plumb archives for the details that turn history into narrative and bring an untold story to life. For the last twelve years, he has been my sounding board, confidant, closest advisor, and partner in all things, and *Hidden Figures* would not have happened without his support. For everything, Aran: my boundless respect, deepest gratitude, and endless love.

NOTES

Page Number

xiv "GS-9 Research Scientist": W. Kemble Johnson to NACA, "Fair Employment," December 14, 1951. National Archives at Philadelphia, hereinafter referred to as NARA Phil.

xiv two white head computers: Blanche S. Fitchett personnel file, US Civil Service Commission. National Personnel Records Center, hereinafter referred to as NPRC.

xiv 'This is a scientist, this is an engineer': *Women Computers*, video recorded at NASA Langley, December 13, 1990. https://www.youtube.com/watch?v=o-MN3Cp2Cpc.

xiv "I just assumed they were all secretaries": Ibid.

xiv Five white women: "What's My Name?" *Air Scoop*, June 14, 1946.

xiv "Several hundred": Beverly Golemba, "Human Computers: The Women in Aeronautical Research," PhD dissertation, St. Leo College, 1994, 4. Available at NASA Cultural Resources, http://crgis.ndc.nasa.gov/crgis/images/c/c7/Golemba.pdf.

xvi "Spacetown USA": James R. Hansen, *Spaceflight Revolution: NASA Langley Research Center from Sputnik to Apollo* (Washington, DC: National Aeronautics and Space Administration, 1995). The town was dubbed "Spacetown, USA" at the October 5, 1962, astronaut parade celebrating Project Mercury. Hansen's book (p. 78–79) has wonderful pictures of the day.

CHAPTER 1: A DOOR OPENS

1 "This establishment has urgent need": Melvin Butler to Chief of Field Operations, Telegram, US Civil Service Commission, May 13, 1943, NARA Phil.

1 Every morning at 7:00 a.m.: M. J. McAuliffe to Recruiting Representatives, Fourth Regional Office, "Recruiting Workers for National Advisory Committee for Aeronautics (Langley Memorial Aeronautical Laboratory)," January 28, 1944, NARA Phil.

1 Dispatching the lab's station wagon: Ibid.

1 *so help me God*: This is the oath taken by all federal civil servants. The full text is available at https://www.law.cornell.edu/uscode/text/5/3331.

2 with a jump seat: W. Kemble Johnson, interview with Michael D. Keller, June 27, 1967, Langley Archives Collection, hereinafter referred to as LAC.

2 500-odd employees: James Hansen, *Engineer in Charge: A History of the Langley Aeronautical Laboratory, 1917–1958* (Washington, DC: National Aeronautics and Space Administration, 1987). Statistics taken from Appendix B, "Growth of Langley Staff, 1919–1958," 413.

3 fifty thousand per year: Franklin D. Roosevelt, Message to Congress on Appropriations for National Defense, May 16, 1940, http://www.presidency.ucsb.edu/ws/?pid=15954

3 ninety planes a month: Arthur Herman, *Freedom's Forge: How American Business Produced Victory in World War II* (New York: Random House Publishing Group, 2012), 11.

3 the largest industry in the world: Judy A. Rumerman, "The American Aerospace Industry During World War II," US Centennial of Flight Commission website, http://www.centennialofflight.net/essay/Aerospace/WWII_Industry/Aero7.htm. Comparative aircraft production statistics at Wikipedia: https://en.wikipedia.org/wiki/World_War_II_aircraft_production.

4 started in 1935: "What's My Name?"

4 investing $500: R. H. Cramer to R. A. Darby, "Computing Groups Organization and Practice at NACA," April 27, 1942, http://crgis.ndc.nasa.gov/crgis/images/7/76/ComputingGroupOrg1942.pdf.

5 grudgingly admitted: Ibid.

5 a boost to the laboratory's bottom line: Ibid.

5 "Reduce your household duties!": February 3, 1942, Langley Archives Collection (hereafter referred to as LAC).

5 "Are there members of your family": "Special Message to the Staff," *Air Scoop*, September 19, 1944.

5 "Who the hell is this guy Randolph?": Jervis Anderson, *A. Philip Randolph: A Biographical Portrait* (Berkeley: University of California Press, 1986), 259.

6 "the stare of an eagle": Ibid.

6 Sherwood's, group had already moved there: NARA Phil.

7 Melvin Butler himself hailed from Portsmouth: Jennifer Vanhoorebeck, "T. M. Butler, Hampton Leader, Dies," *Daily Press*, May 11, 1996.

7 practical solutions: In the NACA's charter was the charge "to supervise and direct the scientific study of the problems of flight with a view to their practical solution." The pragmatic, empirical approach to aeronautical research was one of the agency's defining characteristics and permeated every aspect of its work. For more on the NACA's early days, see Hansen, *Engineer in Charge*, chapter one.

8 bearing the words COLORED GIRLS: Miriam Mann Harris, "Miriam Daniel Mann," September 12, 2011, LAC.

CHAPTER 2: MOBILIZATION

9 100-plus degrees: "The Weather of 1943 in the United States," Monthly Weather Review, December 1943, accessed July 23, 2015, http://docs.lib.noaa .gov/rescue/mwr/071/mwr-071-12-0198.pdf.

9 eighteen thousand bundles of laundry each week: "A Short History of Camp Pickett," Camp Pickett Post Public Information Office, April 1951, 6.

9 stood at the sorting station: Dorothy J. Vaughan Personnel File, US Civil Service, NPRC.

9 the Port of Embarkation: "A Short History of Camp Pickett," 3.

10 stemmers in the tobacco factories: W.E.B. Du Bois, "The Negroes of Farmville, Virginia: A Social Study," *Bulletin of the Department of Labor* 14 (January 1898): 1–38, https://fraser.stlouisfed.org/docs/publications/bls/bls_v03_0014 _1898.pdf.

10 supported three workers: Kathryn Blood, "Negro Women War Workers," Bulletin 205 (Washington, DC: US Department of Labor, Women's Bureau, 1945), 8, http://digitalcollections.smu.edu/cdm/ref/collection/hgp/id/431.

10 40 cents an hour: Vaughan Personnel File.

10 Only a week had elapsed: Ibid.

10 "upper level of training and intelligence in the race": Fred McCuistion, "The South's Negro Teaching Force," *Journal of Negro Education*, April 1932, 18.

10 "direct its thoughts and head its social movements": Ibid.

10 a barbershop, a pool hall, and a service station: Ann Vaughan Hammond, personal interview, April 2, 2014.

10 house on South Main Street: Ibid.

11 ranked in the bottom quarter: Robert Margo, *Race and Schooling in the South, 1880–1950: An Economic History* (Chicago: University of Chicago Press, 1950), 53.

11 almost 50 percent less: Ibid.

11 bested what she earned as a teacher: Vaughan Personnel File.

12 died when she was just two years old: Dewey W. Fox, *A Brief Sketch of the Life of Miss Dorothy L. Johnson* (West Virginia African Methodist Episcopal Sunday School Convention, January 1, 1926), 3.

12 worked as a charwoman: 1910 US Census. 1910; Census Place: Kansas Ward 8, Jackson, Missouri; Roll: T624_787; Page: 16A; Enumeration District: 0099; FHL microfilm: 1374800, Ancestry.com.

12 before she entered school: Fox, *A Brief Sketch*, 3.

12 vaulted her ahead two grades: Ibid.

12 enrolling her in piano lessons: Ibid.

12 a successful Negro restauranteur: Connie Park Rick, *Our Monongalia* (Terra Alta, WV Headline Books, 1999), 106 and 142. John Hunt met Leonard Johnson on a business trip to Kansas City and was so impressed he invited him to work for him in Morgantown. Dorothy's father earned the moniker "Kansas City" Johnson and eventually opened his own restaurant.

12 Beechhurst School: Fox, *A Brief Sketch*, 6.

12 valedictorian's spot: Ibid., 6.

12 "This is the dawn of a life": Ibid., 8.

13 "splendid grades": Ibid., 5.

13 recommended her for graduate study: Ann Vaughan Hammond, personal interview, June 30, 2014.

13 the inaugural class for a master's degree: "Mathematicians of the African Diaspora: Dudley Weldon Woodard," University at Buffalo, State University of New York Mathematics Department, http://www.math.buffalo.edu/mad/PEEPS/woodard_dudleyw.html

13 the first two Negroes to earn doctorates in mathematics: Johnny L. Houston, "Elbert Frank Cox," *National Mathematical Association Newsletter*, Spring 1995, 4.

13 like a third of all Americans: Robert A. Margo, "Employment and Unemployment in the 1930s," *Journal of Economic Perspectives* 7, no. 2 (Spring 1993) 42.

13 felt it was her responsibility: Hammond interview, June 30, 2014.

14 in rural Tamms, Illinois: Vaughan Personnel File.

14 the school ran out of money: Ibid.

14 as an itinerant bellman: Hammond interview, April 2, 2014.

14 at the Greenbrier: Ibid.

15 Dorothy spied the notice: Ibid.

15 "This organization is considering a plan": W. Kemble Johnson to Grace Lawrence, February 5, 1942, and Mary W. Watkins to W. Kemble Johnson, February 9, 1942, NARA Phil.

15 "It is expected that outstanding students": Ibid.

15 "since the Emancipation Proclamation": Jervis Anderson, *A. Philip Randolph: A Biographical Portrait* (Berkeley: University of California Press, 1986), 259.

15 sister-in-law had moved to Washington: Vaughan Personnel File.

16 "Paving the Way for Women Engineers": "Paving the Way for Women Engineers," *Norfolk Journal and Guide*, May 8, 1943.

16 running a nursery school: "Hampton School Head Urges Students to Remain in School," *Norfolk Journal and Guide*, September 4, 1943.

16 Mary Cherry: "Paving the Way for Women Engineers."

16 teaching machine shop: Miriam Mann Harris, personal interview, May 6, 2014.

17 reviewed her qualifications in detail: Vaughan Personnel File.

17 48 hours: Ibid.

CHAPTER 3: PAST IS PROLOGUE

19 to accommodate 180 students: Jarl K. Jackson and Julie L. Vosmik, "National Historic Landmark Nomination: Robert Russa Moton High School," National Park Service, December 1994, https://www.nps.gov/nhl/find/statelists/va/Moton.pdf

19 167 students arrived for classes: Ibid.

19 Farmville chapter of the NAACP: "New NAACP Branches Formed in Two Counties," *Norfolk Journal and Guide*, January 14, 1939.

19 an auditorium outfitted with folding chairs: Bob Smith, *They Closed Their Schools: Prince Edward County, Virginia, 1951–1964* (Chapel Hill: University of North Carolina Press, 1965), 60.

20 vocal quartets had come away victorious: "500 Students in VA State Music Festival," *Norfolk Journal and Guide*, April 20, 1935.

20 "hardest working director": Ibid.

20 "The Light Still Shines": Eloise Barker, "Farmville," *Norfolk Journal and Guide*, December 11, 1943.

20 4-H club made care packages: "Farmville," *Norfolk Journal and Guide*, November 28, 1942.

20 "What Can We Do to Win the War?": Ibid.

20 put war stamps on sale: Ibid.

20 held going-away parties: Patrick Louis Cooney and Henry W. Powell, "Vagabond: 1933–1937," *The Life and Times of the Prophet Vernon Johns: Father of the Civil Rights Movement* (Vernon Johns Society), http://www.vernonjohns.org/tcal001/vjvagbnd.html.

20 a unit called Wartime Mathematics: Vaughan Personnel File; Alan W. Garrett, "Mathematics Education Goes to War: Challenges and Opportunities during the Second World War," paper presented at the Annual Meeting of the National Council of Teachers of Mathematics, April 21, 1999.

21 "pay at the rate of $2,000 per annum": Vaughan Personnel File.

21 $850 annual salary: Ibid.

21 "instructor in mathematics": Barker, "Farmville," December 11, 1943.

21 only until the bell rang at the front door: Hammond interview, June 30, 2014.

22 in 1932 when they married: Hammond interview, April 4, 2014.

23 an evening extension course in education: Vaughan Personnel File.

23 accompanied him to White Sulphur Springs: Hammond interview, June 30, 2014.

23 setting foot on the hotel grounds: Ibid.

23 peering through the shrubbery-covered iron fence: Ibid.

23 German and Japanese detainees: Robert S. Conte, *The History of the Greenbrier: America's Resort* (Parkersburg, WV: Trans Allegheny Books, 1989), 133.

23 an older Negro couple: Katherine Johnson, personal interview, September 17, 2011.

24 graduated from high school at fourteen: Katherine Johnson, personal interview, March 6, 2011.

24 every math course in the school's catalog: Ibid.

24 created advanced math classes: Katherine Johnson, personal interview, September 27, 2013.

24 third Negro in the country: "University History: Pioneer African American Mathematicians," University of Pennsylvania, http://www.archives.upenn.edu /histy/features/aframer/math.html.

24 in 1929: Ibid.

24 degree in math and French: Heather S. Deiss, "Katherine Johnson: A Lifetime of STEM," NASA.gov, November 6, 2013, http://www.nasa.gov/audience /foreducators/a-lifetime-of-stem.html

24 she was denied admission: "Virginia Women in History: Alice Jackson Stuart," Library of Virginia, http://www.lva.virginia.gov/public/vawomen/2012 /?bio=stuart.

24 continued until 1950: Ibid.

25 "unusually capable": Albert P. Kalme, "Racial Desegregation and Integration in American Education: The Case History of West Virginia State College, 1891–1973," PhD dissertation, University of Ottawa, 1976, 149.

25 decided to leave WVU's graduate program: Johnson interview, March 6, 2011.

CHAPTER 4: THE DOUBLE V

27 by the hundreds of thousands: Charles F. Marsh, ed., *The Hampton Roads Communities in World War II* (Chapel Hill: University of North Carolina Press, 1951/2011), 77.

28 the melodies of a hundred different hearts and hometowns: "Hampton Roads Embarkation Series, 1942–1946," US Army Signal Corps Photograph Collection, Library of Virginia digital archive; http://www.lva.virginia.gov/exhibits /treasures/arts/art-m12.htm. All descriptions in this paragraph are taken from photographs in the collection.

28 coverall-clad women: "What's a War Boom Like?" *Business Week*, June 6, 1942, 24.

28   hired women to pose as mannequins: Ibid.

28   exploded from 393,000 to 576,000: Marsh, *The Hampton Roads Communities*, 77.

28   from 15,000 to more than 150,000: Ibid.

28   PLEASE WASH AT HOME: "What's a War Boom Like?" 28.

28   showed movies from 11:00 a.m. to midnight: Ibid.

29   *Victory Through Air Power*: Walt Disney Productions, 1943.

29   still enjoyed a waiting list: "What's a War Boom Like?"; Marsh, *The Hampton Roads Communities*; William Reginald, *The Road to Victory: A History of Hampton Roads Port of Embarkation in World War II* (Newport News, VA: City of Newport News, 1946).

29   5,200 prefabricated demountable homes: "Newsome Park Homes Defense Workers," *Norfolk Journal and Guide*, March 6, 1943.

29   arrived in Newport News on a Thursday: Vaughan Personnel File.

29   "avoid embarrassment": W. Kemble Johnson to Staff, "Living Facilities for New Employees," September 1, 1942, NARA Phil.

29   Five dollars a week: "Local Housing Facilities Available to NACA Employees," January 1944, NARA Phil.

29   Frederick and Annie Lucy: Ann Vaughan Hammond, personal interview, June 30, 2014; 1940 US Census, Ancestry.com.

29   owned a grocery store: Ibid.

30   plans to open the city's first Negro pharmacy: "Smith's Pharmacy," National Register of Historic Places Registration Form, National Park Service, April 18, 2002, http://www.dhr.virginia.gov/registers/Cities/NewportNews/121-5066_ Smiths_Pharmacy_2002_Final_Nomination.pdf

30   Whittaker Memorial opened earlier in 1943: "Whittaker Memorial Hospital," National Register of Historic Places Registration Form, National Park Service, August 19, 2009, http://www.dhr.virginia.gov/registers/Cities/Newport News/121-5072_Whittaker_Memorial_Hospital_2009_FINAL_NR.pdf.

30   Whites entered and exited: Virginius Dabney, "To Lessen Race Friction," *Richmond Times Dispatch*, November 13, 1943; "VPS Begins Two Man Operation," *Norfolk Journal and Guide*, November 14, 1942.

31   wrote a letter to the bus company: Theresa Holloman and Evelyn Fauntleroy, "Local Women Protest Bus Drivers' Discourtesies," *Norfolk Journal and Guide*, June 5, 1943.

31 denied entry to Negro men: "An Investigation Is Indicated Here," *Norfolk Journal and Guide*, March 17, 1945.

31 "Men of every creed": Franklin Delano Roosevelt, *The Four Freedoms: Message to the 77th Congress*, January 6, 1941, http://www.fdrlibrary.marist.edu /pdfs/fftext.pdf.

31 "Four Freedoms": Ibid.

32 "With thousands of your sons in the camps": Herbert Aptheker, "Status of Negroes in Wartime Revealed," *Norfolk Journal and Guide*, April 26, 1941.

32 "I felt damned glad": Genna Rae McNeil, *Groundwork: Charles Hamilton Houston and the Struggle for Civil Rights* (Philadelphia: University of Pennsylvania Press, 1983), 1283.

33 A 1915 rule requiring a photo: Samuel Krislov, *The Negro in Federal Employment* (New Orleans: Quid Pro Quo, 2012).

33 purging the rolls of high-ranking black officials: John A. Davis and Cornelius Golightly, "Negro Employment in the Federal Government," *Phylon*, 1942, 338.

33 "There is no power in the world": John Temple Graves, "The Southern Negro and the War Crisis," *Virginia Quarterly Review*, Autumn 1942.

33 ripped Negroes asunder: W. E. B. Du Bois, *The Souls of Black Folk*, 1903, University of Virginia, http://web.archive.org/web/20081004090243/http://etext .lib.virginia.edu/toc/modeng/public/DubSoul.html.

33 "Every type of brutality perpetrated by the Germans": Cooney and Powell, *The Life and Times of the Prophet Vernon Johns*. http://www.vernonjohns.org /tcal001/vjthelgy.html.

34 "brilliant scholar-preacher": Taylor Branch, *Parting the Waters: America in the King Years* (New York: Simon & Schuster, 2007), 6.

34 "Help us to get some of the blessings of democracy": P. B. Young, "Service or Betrayal?" *Norfolk Journal and Guide*, April 25, 1942.

34 "Being an American of dark complexion": James G. Thompson, "Should I Sacrifice to Live 'Half-American'?" *Pittsburgh Courier*, January 31, 1942.

35 "as surely as the Axis forces": Ibid.

CHAPTER 5: MANIFEST DESTINY

37 "If the Placement Officer shall see fit": "The First Epistle of the NACAites," *Air Scoop*, January 19, 1945.

38   a disproportionate number of Hampton citizens: F. R. Burgess, "Uncle Sam's Eagle's Saved Hampton," *Richmond Times Dispatch*, January 13, 1935.

38   "The future of this favored section of Virginia": Hansen, *Engineer in Charge*, 16.

38   "life-giving energy": Ibid.

40   2 percent of all black women: Blood, *Negro Women War Workers*, 19–23.

40   Exactly zero percent: Ibid.

40   10 percent of white women: US Bureau of the Census 1940 population survey.

40   "best and biggest aeronautical research complex": Hansen, *Engineer in Charge*, 188.

40   after graduating from Idaho State University: Margery E. Hannah Personnel File, US Civil Service Commission, NPRC.

40   the "English critic": Edward Sharp to Staff, "Change in Computers' Telephone Number," July 31, 1935, NARA Phil.

42   "You men and women working here": "Frank Knox Praises NACA," *Air Scoop*, November 6–12, 1943.

42   The employees spread out from one side of the room: All descriptions are from the Langley archive photo L-35045, NASA Cultural Resources, November 4, 1943, http://crgis.ndc.nasa.gov/historic/File:L-35045.jpg.

43   walked over to the cafeteria: "Knox to Visit LMAL Nov. 4," *Air Scoop*, October 30–November 4, 1943. The lab rescheduled the employees' usual lunch times that day in order to accommodate Knox's speech.

43   COLORED COMPUTERS: Miriam Mann Harris, personal interview, May 6, 2014; Miriam Mann Harris, "Miriam Daniel Mann Biography," NASA Cultural Resources, September 12, 2011, http://crgis.ndc.nasa.gov/crgis/images/d/d3/MannBio.pdf.

44   Anne Wythe Hall: "Girls Prepare to Move into Wythe Hall," *Air Scoop*, November 20–26, 1943.

44   "There's my sign for today": Harris interview.

44   banish it to the recesses of her purse: Ibid.

44   Irene Morgan: Derek C. Catsam and Brendan Wolfe, "*Morgan v. Virginia* (1946)," *Encyclopedia Virginia*, October 20, 2014.

45   The NAACP Legal Defense Fund: Richard Goldstein, "Irene Morgan Kirkaldy, 90, Rights Pioneer, Dies," *New York Times*, August 13, 2007.

45   "They are going to *fire* you": Harris interview.

45 former plantation named Shellbanks Farm: Sharon Loury, "Notes from *The Beverley Family of Virginia*," NASA Cultural Resources, 1956, http://crgis.ndc .nasa.gov/crgis/images/9/90/BeverleyFamily.pdf.

45 the sale of the 770-acre property: "Hampton Institute Sells Farm to War Department," *Baltimore Afro-American*, January 4, 1941.

45 one of the largest air bases in the world: Ibid.

45 a thousand black naval recruits: S. A. Haynes, "Navy Officials Praise Work at Hampton Naval Training Station, First of Its Kind," *Norfolk Journal and Guide*, September 11, 1943.

46 Naval Air Station Patuxent River: James A. Johnson, personal interview, June 11, 2011.

46 "the greatest break in history": "Workers in War Industry Discussed in Conference," *Norfolk Journal and Guide*, July 4, 1942.

46 he urged white colleges: "White Colleges Urged to Employ Colored Profs," *Baltimore Afro-American*, May 24, 1941.

46 dance with a Hampton coed: "Dr. MacLean's Resignation Accepted by Hampton Board," *Baltimore Afro-American*, February 6, 1943.

46 corresponding with Orville Wright: H. J. E. Reid's correspondence is almost as interesting a record of local happenings as it is a chronicle of operations at the laboratory. NARA Phil.

46 Kiwanis Club set: "Dr. MacLean's Resignation Accepted by Hampton Board."

46 neither left fingerprints: After six years of research, a formal document that paved the way for the establishment of the West Computing office remains elusive. Given the need to establish a segregated office and separate bathrooms for the black women, and given the customs of the time and place, this certainly seems to be the kind of decision that would have required knowledge and sign-off—from the top. But after scouring MacLean's papers at Hampton Institute, reviewing the FEPC documents from his time as the head of the committee, poring over NASA and Langley archives at the Langley Research Center and NASA headquarters, examining Reid's correspondence and Fair Employment files at NARA Philadelphia, going through the wartime records of the Education Department (which oversaw the ESMWT), and the civil service and War Manpower Commission records (at NARA College Park and NARA Philadelphia, respectively) I am led to conclude that this was a handshake deal.

47   the world was coming to an end: *Women Computers*.

47   made Marge a pariah: Katherine Johnson, interview with Aaron Gillette, September 17, 1992, NASA HQ.

47   harassing a black man: Dave Lawrence, "Langley Engineer Is Remembered for Part in History," *Daily Press*, August 21, 1999.

47   Arthur Kantrowitz, bailed Jones out: Ibid.

48   purchase of war bonds: Each weekly issue of *Air Scoop*, from 1942 through 1945, tallied war bond purchases by group; West Computing was routinely at the top of the list.

CHAPTER 6: WAR BIRDS

51   *Flyers Help Smash Nazis!*: *Norfolk Journal and Guide*, May 27, 1944.

51   The "Tan Yanks": John Jordan, "Negro Pilots Sink Nazi Warship," *Norfolk Journal and Guide*, July 8, 1944.

51   flying North American P-51 Mustangs: Ibid.

52   "as the war enters its decisive stage": "Missions Take Fliers into Five Countries," *Norfolk Journal and Guide*, July 15, 1944.

52   "a 'pilot's airplane' ": "New US 'Mustang' Heralded as Best Fighter Plane of 1943," *Washington Post*, November 27, 1942.

52   "I will get you up in the air": "Tuskegee Airman Reunites with 'Best Plane in the World,' " NASA, June 10, 2004, http://www.nasa.gov/vision/earth/improvingflight/tuskegee.html.

52   "Laboratories at war!": "Cites Importance of Research in War Effort," *Air Scoop*, March 25–31, 1944.

52   "You tell it to someone": *Air Scoop*, March 25–31, 1944.

53   nearly lost her raccoon coat: Hansen, *Engineer in Charge*, 254.

53   "New York Jews": Pearl I. Young, interview with Michael D. Keller, January 10, 1966, LAC.

53   "weirdos": Parke Rouse, "Early Days at Langley Were Colorful," *Daily Press*, March 25, 1990.

53   dismantling a toaster: Milton A. Silveira, interview with Sandra Johnson, JSC, October 5, 2005.

53   with books on their steering wheels: *Women Computers*.

53   as a runway: Golemba, "Human Computers," 37.

54 best engineering graduate school program in the world: Alex Roland, *Model Research: The National Advisory Committee for Aeronautics 1915–1958* (Washington, DC: NASA, 1985), 275.

54 for new computers: "Computers Attend Physics Classes," *LMAL Bulletin*, June 28, 1943.

54 weekly two-hour laboratory session: Ibid.

54 four hours of homework: Ibid.

54 men such as Arthur Kantrowitz: Ibid.

55 P-51 Mustang was the first production plane: Hansen, *Engineer in Charge*, 116.

55 Ann Baumgartner Carl: Katherine Calos, "Ann G. B. Carl, First US woman to Fly Jet, Dies," *Richmond Times-Dispatch*, March 22, 2008.

55 "damn fool's job": "Transport: Damn Fool's Job," *Time*, April 1, 1935.

56 No organization came close to Langley: Hansen, *Engineer in Charge*, 46.

57 the lab's Flight Research Division: Fitchett Personnel File.

58 results and recommendations of the NACA: "We Backed the Attack," *LMAL Bulletin*, June 24–30, 1944.

58 "a cut above": Sugenia M. Johnson, interview with Rebecca Wright, JSC, April 2, 2014.

58 "Woe unto thee": "Second Epistle of the NACAites," *Air Scoop*, January 26, 1945.

59 "final bombing of Japan": "We Backed the Attack."

CHAPTER 7: THE DURATION

61 She signed a lease: K. Elizabeth Paige, "Newsome Park Echoes," *Norfolk Journal and Guide*, July 8, 1944.

61 Protective paper . . . covered the floors: *Newsome Park Reunion: The Legacy of a Village*, event program, September 6, 2006, 6, in author's possession.

62 stay with her during a school break: Hammond interview, June 30, 2014.

62 Aberdeen Gardens, a Depression-era subdivision: "Aberdeen Gardens," National Register of Historic Places Registration Form, National Park Service, March 7, 1944, http://www.dhr.virginia.gov/registers/Cities/Hampton/114 -0146_Aberdeen_Gardens_HD_1994_Final_Nomination.pdf.

62 440 acres: Ibid.

62 "high type suburban community for Negro families": W. R. Walker Jr., "Mimosa Crescent, Post-War Housing Project, Started," *Norfolk Journal and Guide*, July 15, 1944.

63 peddling their wares to the neighbors: Catherine R. Weaver, "Memories of the Village," *Newsome Park Reunion*, event program, September 3, 2005, 6, in author's possession.

64 flooded "joyous tumult": C. I. Wiliams, "City Greets Victory With Joyous Tumult," *Norfolk Journal and Guide*, August 19, 1945.

65 "indescribable noise-making devices": Ibid.

65 "It seems impossible to escape the conclusion": "Hampton Roads Area Faces Drastic Cut in Employment," *Washington Post*, October 21, 1945.

66 their white, Gentile-only employment policies: "Jobs Open for Whites Only," *Norfolk Journal and Guide*, September 1, 1945.

66 "the most dangerous idea ever seriously considered": Glenn Feldman, *The Great Melding: War, the Dixiecrat Rebellion, and the Southern Model for America's New Conservatism* (Tuscaloosa: University of Alabama Press, 2015), 211.

66 "following the Communists' lead": Ibid., 299.

66 "the most urbane and genteel dictatorship in America": John Gunther, *Inside USA.* (New York: Harper and Brothers, 1947), 705.

66 helped fellow Virginian Woodrow Wilson win the White House in 1912: Ronald L. Heinmann, "The Byrd Legacy: Integrity, Honesty, Lack of Imagination, Massive Resistance," *Richmond Times-Dispatch*, August 25, 2013.

67 "war-devastated populations in Europe": "Realtors Win Efforts for Post-war Riddance of Federal Housing Units," *Norfolk Journal and Guide*, June 30, 1945.

67 "not permanent in its current location": Ibid.

67 room and board to a returning military man: Hammond interview, June 30, 2014.

67 hosting a party for nearly twenty people: K. Elizabeth Paige, "Newsome Park Echoes," *Norfolk Journal and Guide*, September 30, 1944.

CHAPTER 8: THOSE WHO MOVE FORWARD

69 at least eight people in Smyth County: Katherine Johnson, personal interview, March 13, 2011.

69 careful to keep her teaching certificate current: Ibid.

69 "If you can play the piano": Ibid.

69 were ordered to move to the back: "Katherine Johnson, National Visionary," National Visionary Leadership Project, http://www.visionaryproject.org/johnsonkatherine/.

70 evicted the black passengers: Ibid.

70 Katherine earned $50 a month: Johnson interview, March 13, 2011.

70 less money than the school's white janitor: Mark St. John Erickson, "No Easy Journey," *Daily Press*, May 1, 2004.

70 when a $110-a-month job offer: Johnson interview, March 13, 2011.

70 "and no one is better than you": Johnson interview, December 27, 2010.

71 how many board feet a tree would yield: "What Matters—Katherine Johnson: NASA Pioneer and 'Computer,'" WHRO Television Broadcast, February 25, 2011, https://www.youtube.com/watch?v=r8gJqKyIGhE.

71 skipped ahead from second grade to fifth: "Katherine Johnson, National Visionary."

71 they'd find their pupil in the classroom next door: Johnson interview, December 27, 2010.

71 Joseph and Rose Kennedy: Conte, *The History of the Greenbrier*, 113.

71 Bing Crosby, the duke of Windsor: Ibid., 148–49.

72 segmented its serving class: Robert S. Comte, personal interview, September 12, 2012.

72 "Tu m'entends tout, n'est-ce pas?" Johnson interview, December 27, 2010.

72 The Greenbrier's Parisian chef: Ibid.

72 taught him Roman numerals: Ibid.

73 served as the college's dean: Lorenzo J. Greene and Arvarh E. Strickland, *Selling Black History for Carter G. Woodson: A Diary* (Columbia: University of Missouri Press, 1996), 194.

73 Civilian Aide in the War Department: "College and School News," *The Crisis*, January 1944; "James C. Evans Dies," *Washington Post*, April 17, 1988.

73 and a mean game of tennis: Johnson interview, March 6, 2011.

73 flew so low over the house of the school's president: Margaret Claytor Woodbury and Ruth C. Marsh, *Virginia Kaleidoscope: The Claytor Family of Roanoke, and Some of Its Kinships, from First Families of Virginia and Their Former Slaves* (Ruth C. Marsh, 1994), 202.

73 his drawling "country" accent: Ibid.

73 furiously scribbled mathematical formulas on the chalkboard: Ibid.

73 "You would make a good research mathematician": Johnson interview, March 11, 2011.

73 received an offer to join the inaugural class: Hammond interview, June 30, 2014.

73 a significant advance in the field: "Pioneer African American Mathematicians," University of Pennsylvania, http://www.archives.upenn.edu/histy/features /aframer/math.html.

74 "If young colored men receive scientific training": W. E. B. Dubois, "The Negro Scientist," *The American Scholar* 8, no. 3 (Summer 1939): 316.

74 "The [white] libraries": Ibid.

74 "no opportunity to go to scientific meetings": Jacqueline Giles-Girron, "Black Pioneers in Mathematics: Brown [*sic*], Granville, Cox, Claytor and Blackwell," *Focus: the Newsletter of the Mathematical Association of America* 11, no. 1 (January–February 1991): 18.

74 just over a hundred women: Margaret Rossiter, *Women Scientists in America: Before Affirmative Action 1940–1972* (Baltimore: Johns Hopkins University Press, 1995), 137.

74 Irish and Jewish women with math degrees: David Alan Grier, *When Computers Were Human* (Princeton, NJ: Princeton University Press, 1997), 208–9.

74 "But where will I find a job?" Johnson interview, December 27, 2010.

74 they got married, telling no one: Johnson interview, March 13, 2011.

74 waiting outside her classroom: Johnson interview, September 27, 2013.

75 walked away from an offer of $4 million: Albert P. Kalme, "Racial Desegregation and Integration and American Education: The Case History of West Virginia State College, 1891–1973," PhD dissertation, University of Ottawa, 1973, 173.

75 "So I picked you": Johnson interview, September 27, 2013.

75 presented her with a full set of math reference books: Ibid.

76 they were expecting their first child: Ibid.

CHAPTER 9: BREAKING BARRIERS

78 had it all figured out: Leonard Vaughan, personal interview, April 3, 2014.

78 Howard Vaughan's sister-in-law: Hammond interview, April 2, 2014; Joanne Cavanaugh Simpson, "Sound Reasoning," *Johns Hopkins Magazine*, September 2003.

78 "for members of the race": "New Peninsula Beach Opens Memorial Day," *Norfolk Journal and Guide*, May 27, 1944.

78 spent weeks organizing the menu: Harris interview.

78 roasting marshmallows over a fire: Ibid.

78 founded the resort in 1898: Mark St. John Erickson, "Remembering One of the South's Premier Black Seaside Resorts," *Daily Press*, August 21, 2013.

79 $2,000 a year: Vaughan Personnel File.

79 just $96: Martha J. Bailey and William J. Collins, "The Wage Gains of African-American Women in the 1940s," *Journal of Economic History* 66, no. 3 (September 2006): 737–77.

79 took a walk around the block: Michelle Webb, personal interview, February 10, 2016.

79 headquarters of its Tactical Air Command: "Gen. Devers Takes Command of Fort Monroe, New AGF Base," *Washington Post*, October 2, 1946.

80 "military-industrial complex": Dwight D. Eisenhower, "Farewell Address," January 17, 1961, https://www.ourdocuments.gov/doc.php?doc=90&page=transcript.

80 more than three thousand employees: Hansen, *Engineer in Charge*, 413.

80 tendered their resignations: Golemba, "Human Computers," 90.

81 top-ranked managers: Ibid., 90–91.

81 "excellent" ratings: Ibid.

81 had been appointed shift supervisors: Fitchett Personnel File.

81 had swelled to twenty-five women: Ibid.

81 often worked the 3:00 p.m.-to-11:00 p.m. shift: Golemba, "Human Computers," 87.

81 "two spacious offices": "Blanche Sponsler Called in . . . ,"*Air Scoop*, August 24, 1945.

81 vacancies at the laboratory: "Vacancies Open Here at Lab," *Air Scoop*, August 9, 1946.

82 "Cadettes": The Cadettes were formed after the Langley laboratory recommended that Curtiss Wright adopt its female computing pool setup, as detailed in R. H. Cramer's April 27, 1942, memo "Computing Groups Organization and Practice at NACA" (see LAC). Natalia Holt's book *Rise of the Rocket Girls: The Women Who Propelled Us, from Missiles to the Moon to Mars* (New York: Little, Brown, 2016), David Alan Grier's *When Computers Were Human* (Princeton, NJ: Princeton University Press, 2005), and Margaret Ros-

siter's *Women Scientists in America: Before Affirmative Action 1940–1972* (Baltimore: Johns Hopkins University Press, 1995) all offer fascinating accounts of the computers who worked at installations other than the NACA.

82 subprofessional scientific aide: Walter T. Vicenti, *Robert Thomas Jones 1910– 1999: A Biographical Memoir* (Washington, DC: National Academies Press, 2005).

83 conspired to skip him ahead to a P-2: William R. Sears, "Introduction," *Collected Works of R. T. Jones* (Moffett Field, CA: National Aeronautics and Space Administration, 1976), ix.

83 lunchtime conversations: John V. Becker, *The High Speed Frontier: Case Histories of Four NACA Programs, 1920–1950* (Washington, DC: NASA, 1980), 14.

83 men-only smokers: Edward R. Sharp, "Smoker for Men Only," Memorandum for Section Heads and Division Chiefs, November 26, 1935, NARA Phil.

84 including two former East Computers: Sheryll Goecke Powers, *Women in Flight Research at NASA Dryden Flight Research Center from 1946 to 1995* (Washington, DC: National Aeronautics and Space Administration, 1997), 3.

85 corroborated by the female computers on the ground: Ibid., 12.

85 NACA's lone female author: After an extensive name search of NASA Technical Reports Server (NTRS) and scouring the references in other NACA reports published in the 1930s, 1940s, and 1950s, Doris Cohen's was the only female name I could find until the mid-1940s, when the names of other women began to appear on the publications. Her name first appeared along with Robert T. Jones on "An Analysis of the Stability of an Airplane with Free Controls," Langley Aeronautical Laboratory, January 1941, NTRS.

85 Doris Cohen published nine reports: Ibid.

85 (whom she would eventually marry): David F. Salisbury, "Aerodynamics Pioneer R. T. Jones, Former Consulting Professor, Dies," *Stanford University News Service*, August 24, 1999. Their professional-personal partnership was fruitful, culminating in publication of the classic aerodynamics text *High Speed Wing Theory* (Princeton, NJ: Princeton University Press, 1960).

86 four hundred Langley computers received training on Tucker's watch: "What's My Name?"

86 In 1947, the laboratory disbanded East Computing: Floyd L. Thompson to All Concerned, "Disbanding of East Area Computing Pool," September 17, 1947, NASA Phil.

86 She accepted a job at the Northrup Corporation: "Early Alumni and STEM Fields: Virginia Tucker," UNCG Special Collections and University Archives, October 14, 2014, http://uncgarchives.tumblr.com/post/100014384990/early-alumni-and-stem-fields-virginia-tucker.

87 When three West Computers made the leap: *Women Computers*.

87 hadn't even known: Golemba, "Human Computers," 14.

87 Arkansas, Georgia, and Tennessee: Lisa Frazier, "Searching for Dorothy," *Washington Post*, May 7, 2000.

87 P-1 mathematician: Dorothy Hoover Personnel File, US Civil Service Commission, NPRC.

87 start the process of inputting values: Sugenia Johnson interview.

88 most respected analysts: Becker, *The High Speed Frontier*, 14.

88 listening to classical music and discussing politics: Robert A. Bell, "Former 'Discussion Groups' at the NACA Langley Aeronautical Laboratory," Memorandum for the Security Officer, NACA, July 23, 1954, Federal Bureau of Investigation (FBI), https://vault.fbi.gov/rosenberg-case/julius-rosenberg/julius-rosenberg-part-72-of-1.

88 directly for him: Hoover Personnel File.

88 publishing a study with S. Katzoff and Margery E. Hannah, "Calculation of Tunnel-Induced Upwash Velocities for Swept and Yawed Wings," Langley Aeronautical Laboratory, 1948, NTRS.

88 thirty-five-year-old newlywed: *Air Scoop*, October 24, 1947.

89 requested a transfer to the Ames Laboratory: Fitchett Personnel File.

89 one-month illness: Ibid.

90 July and August 1948: Ibid.

90 made an urgent call to Eldridge Derring: Ibid.

90 Blanche had been acting strangely: Ibid.

90 "behaving irrationally": Ibid.

90 Derring, along with the lab's health officer: Ibid.

90 anxiously waiting in the lobby: Ibid.

90 "meaningless words and symbols": Ibid.

90 "I'm trying to explain how to go": Ibid.

90 "0 ±1 to three significant figures": Ibid.

90 "one P-75,000": Ibid.

90 "as some college students": Ibid.

90 "at least four strong men": Ibid.

91 taken away to the Tucker Sanatorium: Ibid.

91 "It appears that she will continue ill indefinitely": Ibid.

91 obituary in the *Daily Press*: Blanche Sponsler Fitchett Obituary, *Daily Press*, May 31, 1949.

91 "dementia praecox": Blanche Sponsler Fitchett death certificate, State of Virginia, May 29 1949, Ancestry.com.

91 appointed Dorothy Vaughan acting head of West Computing: Eldridge H. Derring to All Concerned, "Change in Organization of Research Services and Control," April 12, 1949, NARA Phil.

92 two years to earn the full title of section head: Eldridge H. Derring to All Concerned: "Appointment of Head of West Area Computers Unit," January 8, 1951, NARA Phil.

92 "Effective this date": Ibid.

CHAPTER 10: HOME BY THE SEA

93 the work songs of the black women: Chauncey E. Brown, personal interview, July 19, 2014; *Virginia Traditions, Virginia Work Songs* (Ferrum, VA: Blue Ridge Institute of Ferrum College, 1983).

94 "Confederate-set inferno": Mark St. John Erickson, "The Night They Burned Old Hampton Down," *Daily Press*, August 7, 2013.

94 "educated young people": Robert F. Engs, *Freedom's First Generation: Black Hampton, Virginia 1861–1890* (New York: Fordham University Press, 2004), 158.

94 Mary graduated in 1938 with highest honors: "Mary W. Jackson Federal Women's Program Coordinator," LHA, October 1979, http://crgis.ndc.nasa.gov /crgis/images/9/96/MaryJackson1.pdf.

95 mathematics and physical science: Ibid.

95 two of her sisters: Golemba, "Human Computers," 40.

95 college typing course: Mary W. Jackson Personnel File, US Civil Service Commission, NPRC.

95 welcomed guests at the club's front door: Ibid.

95 played the piano: "Hampton USO Club Activities," *Norfolk Journal and Guide*, May 30, 1942.

96 "sharing and caring": Mary Winston Jackson funeral program, 2005, in author's possession.

96 "a pillar": "Hamptonian Observes 75th Birthday," *Norfolk Journal and Guide*, September 7, 1946.

96 one thousand hours of meritorious service: "Bethel AME Rites Held for Mrs. Emily Winston," *Norfolk Journal and Guide*, December 29, 1962.

96 white dress with black sequins: "USO Secretary Weds Navy Man," *Norfolk Journal and Guide*, November 25, 1944.

97 children of domestic servants, crab pickers, laborers: Janice Johnson, personal interview, April 3, 2014.

97 and steering them toward college: Ibid.

97 three-mile "country" hikes: "Hampton Happenings," *Norfolk Journal and Guide*, October 29, 1949.

97 field trips to the crab factory: Janice Johnson interview.

97 tea at the Hampton Institute Mansion House: "Hostess to Girl Scout Troop," *Norfolk Journal and Guide*, March 14, 1953; Janice Johnson interview.

97 students from the school's Home Economics Department: Janice Johnson interview.

98 Mary was leading her charges: Ibid.

98 That day, however, the lyrics: Ibid.

98 "Hold on a minute!" Ibid.

98 "We are never going to sing this again": Ibid.

98 was required to get a secret security clearance: Jackson Personnel File.

99 an atomic attack: A. B. Chatham, "Dissemination of Combat Information," Office Chief of Army Field Forces, Fort Monroe, Virginia, August 29, 1952, http://koreanwar-educator.org/topics/reports/after_action/combat_information_bulletins/combat_information_bulletins_520829_350_05_56.pdf.

99 "too fast to be identified": Stephen Joiner, "The Jet That Shocked the West," *Air and Space Magazine*, December 2013.

99 "Russia Said to Have Fastest Fighter Plane": Leon Schloss, writing in the *Norfolk Journal and Guide*, February 18, 1950.

99 Building 1244, the largest structure of its kind: Photo caption, *Air Scoop*, March 16, 1951.

100 scored a Collier Trophy: "Collier 1940–1949 Recipients," National Aeronautic Association, https://naa.aero/awards/awards-and-trophies/collier-trophy/collier-1940-1949-winners.

100 Project 506: Robert C. Moyer and Mary E. Gainer, "Chasing Theory to the Edge of Space: The Development of the X-15 at NACA Langley Aeronautical Laboratory," *Quest: The History of Spaceflight Quarterly* 19, no. 2 (2012): 5.

100 close to Mach 7: Ibid.

100 Gas Dynamics Laboratory: "1247 Hypersonic Facilities Complex," NASA Cultural Resources http://crgis.ndc.nasa.gov/historic/1247_Hypersonic_Facilities_Complex. Completed in 1952, the laboratory's name was changed to Hypersonic Facilities Complex.

100 up to Mach 18: Ibid.

101 handed down a death sentence against Ethel and Julius Rosenberg: William R. Conklin, "Atom Spy Couple Sentenced to Die," *New York Times*, April 6, 1951.

101 *How to Spot a Communist*: *How to Spot a Communist*, Armed Forces Information Film no.5, 1950.

101 "who don't show their real faces": Ibid.

101 accused of stealing classified NACA documents: Ronald Radosh and Joyce Milton, *The Rosenberg File* (New Haven, CT: Yale University Press, 1997), 300.

101 nuclear-powered airplane: Ibid.

101 high-speed NACA airfoil: Ibid., 299.

101 based on NACA designs: Ibid.

101 ringing the doorbell in the evenings: Sugenia Johnson interview. Joanne Cavanaugh Simpson, "Sound Reasoning," *Hopkins Magazine*, September 2003.

102 Eastman Jacobs, known for his left-leaning sympathies: Interview with Ira H. Abbott, October 27, 1971. LHA.

102 hours questioning Pearl Young: Pearl Young interview.

102 "New York communist people": Ibid.

102 "practically impossible New York Jews": Ibid.

102 caused a scandal: Ibid.

102 a "black computer": Sugenia Johnson interview.

102 *Air Scoop* published a long list of organizations: "List of groups compiled in Connection with Employees Loyalty Program," *Air Scoop*, October 26, 1951.

103 denied service to the Haitian secretary of agriculture: Mary Dudziak, *Cold War Civil Rights: Race and the Image of American Democracy* (Princeton, NJ: Princeton University Press, 2007), 871.

103 Mahatma Gandhi's personal doctor: Ibid., 878.

104 "Untouchability Banished in India: Worshipped in America": Ibid., 755.

104 At the start of the Korean War: "The Beginnings of a New Era for African Americans in the Armed Services," State of New Jersey, http://www.nj.gov /military/korea/factsheets/afroamer.html.

104 were called up: "Tan Yanks Face Action in Korea," *Norfolk Journal and Guide*, July 8, 1950.

104 "The laboratory has one work unit composed entirely of Negro women": Johnson, "Fair Employment."

105 science textbooks and racial harmony: Walter McDougall, *The Heavens and the Earth: A Political History of the Space Age* (Baltimore: Johns Hopkins University Press, 1997), 8.

105 Christine Richie: Christine Richie, personal interview, July 20, 2014.

105 through the college grapevine: Elizabeth Kittrell Taylor, personal interview, July 12, 2014.

CHAPTER 11: THE AREA RULE

108 alongside several white computers: Richard Stradling, "Retired Engineer Remembers Segregated Langley," *Daily Press*, February 8, 1998.

108 "Can you direct me to the bathroom?": Ibid.

109 native of New Bedford: "14 Receive Service Emblems," *Air Scoop*, December 3, 1954.

109 maintained an office in the Aircraft Loads Building: Langley Aeronautical Laboratory Telephone Directory, LHA, 1949.

109 "dead-weight of social degradation": W. E. B. Dubois, *The Souls of Black Folk* (Chicago: A. C. McClurg and Co., 1903).

109 the American dilemma: In 1944, the Carnegie Foundation funded a groundbreaking, comprehensive report on the state of black America, entitled *An American Dilemma: The Negro Problem and Modern Democracy* (New York: The MacMillan Company, 1946). Its author, Swedish economist Gunnar Myrdal, pointed out the brutal circularity of a system that discriminated

against blacks in virtually every aspect of their lives, then excoriated them when they failed to meet the marks set by whites.

109 answered Czarnecki's greeting with a Mach 2 blowdown: Stradling, "Retired Engineer Remembers Segregated Langley."

110 "Why don't you come work for me?": Ibid.

110 Ray Wright had the intuition: Donald D. Baals and William R. Corliss, *Wind Tunnels of NASA* (Washington, DC: NASA History Office, 1981), 61.

110 "long sought technical prize": Ibid., 61.

110 as much as 25 percent: Richard Whitcomb's Discovery: *Richard Whitcomb's Discovery: The Story of the Area Rule,* video, NASA Langley CRGIS, https://www.youtube.com/watch?v=xZWBVgL8I54.

111 "wasp-waisted": "Air Scientist Whitcomb Cited for 'Wasp-Waist' Theory," *Richmond News Leader,* November 29, 1955.

111 a sit-down with CBS news anchor Walter Cronkite: "Interview Set for Whitcomb with Cronkite," *Daily Press,* October 15, 1955.

111 "Hampton Engineer Besieged by Public": *Daily Press,* October 9, 1955.

111 "an oil refinery under a roof": Baals and Corliss, *Wind Tunnels of NASA,* 71.

112 aeronautical research scientist, graded GS-9: Hoover Personnel file.

112 publication of two reports: Frank Malvestuto Jr. and Dorothy M. Hoover, "Supersonic Lift and Pitching Moment of Thin Sweptback Tapered Wings Produced by Constant Vertical Acceleration," Langley Aeronautical Laboratory, March 1951, http://ntrs.nasa.gov/archive/nasa/casi.ntrs.nasa.gov/199 30082993.pdf; Frank Malvestuto Jr. and Dorothy M. Hoover, "Lift and Pitching Derivatives of Thin Sweptback Tapered Wings with Streamwise Tips and Subsonic Leading Edges at Supersonic Speeds," Langley Aeronautical Laboratory, February 1951, http://ntrs.nasa.gov/archive/nasa/casi.ntrs.nasa.gov/19930082953.pdf.

113 Mary had done since graduating: "Mary W. Jackson, Federal Women's Program Manager," October 1979.

113 Mary Jackson had met Jim Williams: Julia G. Williams, personal interview, July 20, 2014.

113 wary of moving: Ibid.

113 Williams wasn't the first black engineer: Ibid.

113 Several white supervisors refused him: Norman Tippens, "Tuskegee Airman James L. 'Jim' Williams, 77," *Daily Press,* January 23, 2004; Williams interview.

113 raised his hand right away: Williams interview.

113 "Jaybird was as fair as it got": Ibid.

113 "So how long do you think you're going to be able to hang on?": Ibid.

114 given an assignment by John Becker: Golemba, "Human Computers," 64; Langley Memorial Laboratory Telephone Directory, 1952.

114 insisting that her calculations were wrong: Ibid.

114 the problem wasn't with her output: Ibid.

114 John Becker apologized to Mary Jackson: Ibid.

115 It was a cause for quiet celebration: Ibid.

CHAPTER 12: SERENDIPITY

117 little sister Patricia: *Katherine Johnson: Becoming a NASA Mathematician*, Leadership Project, March 8, 2010, https://www.youtube.com/watch?list=PLCwE4GdJdVRLOEyW4PhypNnZIJbYLRTVd&v=jUsyYvrz2qQh ttp://www.visionaryproject.org/johnsonkatherine/.

117 vivacious college beauty queen: "Miss Goble Is Bride of Cpl. Kane Jr.," *Norfolk Journal and Guide*, August 30, 1952. All details of the bride and groom's attire, wedding decoration, and honeymoon plans are from this article.

117 still lived in Marion: "Marion, VA Couple Observes Golden Wedding Anniversary," *Norfolk Journal and Guide*, September 19, 1953.

118 "Why don't ya'll come home with us too?": "Katherine Johnson: Becoming a NASA Mathematician."

118 "I can get Snook a job at the shipyard": Ibid.

118 the director of the Newsome Park Community Center: *Katherine Johnson: Becoming a NASA Mathematician*.

118 coordinated community activities: "Newsome Park Community Center Dedicatorial Exercises Held," *Norfolk Journal and Guide*, July 21, 1945.

119 stayed up nights making school outfits: Johnson interview, March 6, 2011.

119 live-in help: Joylette Hylick Goble, personal interview, October 10, 2011.

120 a painter's job at the Newport News shipyard: Johnson interview, September 17, 2011.

120 the club's assistant director: "Joins USO Staff," *Norfolk Journal and Guide*, May 9, 1953.

120 nine-year West Computing veteran: "Peninsula Spotlight," *Norfolk Journal and Guide*, February 5, 1949.

121 a routine that would persist for the next three decades: Johnson interview, August 27, 2013.

121 modest job rating of SP-3: Katherine Johnson, interview with Aaron Gillette, September 17, 1992.

121 "Don't come in here in two weeks": Ibid.

121 "very, very fortunate": Ibid.

121 three times her salary: Johnson interview, September 17, 2011.

122 initiating a quiet conversation: Johnson interview, September 27, 2013.

122 "The Flight Research Division is requesting two new computers": Ibid.

122 her new deskmates, John Mayer, Carl Huss, and Harold Hamer: John Mayer, Carl Huss, and Harold Hamer, "Investigation of the Use of Controls During Service Operations of Fighter Airplanes," NACA Conference on Aircraft Loads, Flutter and Structures, March 2–4, 1953, Langley Aeronautical Laboratory.

122 "picked up and went right over": Johnson interview, September 27, 2013.

122 the division chief, Henry Pearson: Johnson interview, September 17, 1992; Langley Aeronautical Laboratory Telephone Directory, 1952.

123 got up, and walked way: Johnson interview, September 17, 1992.

124 or she could assume: Ibid.

124 became fast friends: Ibid.

CHAPTER 13: TURBULENCE

125 from the entry level of SP-3 to SP-5: Johnson interview, September 17, 1992.

126 Dorothy was drafted as a consultant: "Computers Help Compile Handbook," *Air Scoop*, August 17, 1951.

127 "black-haired, leather-faced, crew-haircutted human cyclone": "Spotlite by K-P," *LMAL Bulletin*, November 30, 1942.

127 The Maneuver Loads Branch conducted research: A great background in the work being done by these groups at the time can be found in W. Hewitt Phillips, *A Journey in Aeronautical Research: A Career at NASA Langley Research Center* (Washington, DC National Aeronautics and Space Administration, 1998).

128 One of the first assignments: *Katherine Johnson: Becoming a NASA Mathematician.*

128 into the trailing wake of a larger plane: Ibid.

128 they were fascinated: Ibid.

128 as long as half an hour: Christopher C. Kraft Jr., "Flight Measurements of the Velocity Distribution and Persistence of the Trailing Vortices of an Airplane," Langley Aeronautical Laboratory, March 1955, NTRS.

129 "one of the most interesting things she had ever read": *Katherine Johnson: Becoming a NASA Mathematician.*

129 Langley's "Skychicks": *Women Computers.*

129 didn't even realize the bathrooms were segregated: Johnson interview, September 17, 1992.

130 refused to so much as enter the Colored bathrooms: Johnson interview, September 17, 1992.

130 bring a bag lunch and eat at her desk: Johnson interview , March 6, 2011.

130 temptation of the ice cream: Ibid.

130 "by the book": Johnson interview, September 27, 2013.

130 perused *Aviation Week*: Ibid.

131 some of the black employees: This is a subject that came up more than once during interviews with people who have known her.

131 a black or a white roommate: Katherine Goble Moore, personal interview, July 31, 2014.

131 before coming up with a yes: Johnson interview, September 15, 2015.

131 "I want to move our girls out of the projects": Moore interview.

131 the federal Housing and Home Finance Agency: "Government Suspends Demolition," *Norfolk Journal and Guide*, August 26, 1950.

132 Gayle Street, a cul-de-sac: Colita Nichols Fairfax, *Hampton, Virginia* (Charleston, VA: Arcadia Publishing, 2005), 69. This provides good background on Hampton's many black neighborhoods.

133 Mimosa Crescent had expanded: "Mimosa Crescent Project Expanded," *Norfolk Journal and Guide*, March 23, 1946.

133 would even get her own bedroom: Moore interview.

133 located at the base of his skull: Johnson interview, March 13, 2011.

133 James Francis Goble died: "Funeral Services Held for James F. Goble," *Norfolk Journal and Guide*, December 29, 1956.

134 "It is very important": Moore interview.

134 "You will have my clothes ironed": Ibid.

135 Family lore had it: Hylick interview.

135 would recall their grandfather saying: Ibid.

136 had a long conversation about it: Johnson interview, December 27, 2010.

CHAPTER 14: ANGLE OF ATTACK

137 the American Century: Henry R. Luce, "The American Century," *Life*, February 17, 1941, 61–65. Luce, the publisher and founder of *Time* and *Life* magazines, penned this influential editorial in February 1941, urging a conflicted America to take a decisive position in World War II and claim its rightful place of power on the world stage. "The world of the 20th Century, if it is to come to life in any nobility of health and vigor, must be to a significant degree an American Century."

137 bought an "electronic calculator": "Announce New Research Device," *Air Scoop*, March 28, 1947.

137 as many as thirty-five variables: Ibid.

138 would cause an error in all the others: Ibid.

138 easily take a month to complete: Ibid.

138 in a few hours: Ibid.

138 two seconds per operation: Ibid.

138 The whole building shook: Eldon Kordes, interview with Rebecca Wright, JSC, February 19, 2015.

138 then an IBM 650: Theresa Overall, "Mom and IBM," personal blog, February 15, 2014.

138 destined for the lab's finance department: Kordes interview.

138 "Let's run it again!": Ibid.

139 Dorothy wasted no time enrolling: Ann Vaughan Hammond, untitled biographical sketch of Dorothy Vaughan, undated, in author's possession.

140 (the students called them "chicken coops"): Teri Kanefield, *The Girl from the Tar Paper Shacks School: Barbara Rose Johns and the Advent of the Civil Rights Movement* (New York: Harry L. Abrams, 2014).

141 "Not Willing to Wait": "Not Willing To Wait: NAACP Leaders Want Integration 'Now!'," *Norfolk Journal and Guide*, May 29, 1954.

141 "If we can organize the Southern States for massive resistance": Benjamin Muse, *Virginia's Massive Resistance* (Bloomington: Indiana University Press, 1956), 22.

141 some of the black employees attended: Johnson, "Fair Employment."

141 bookkeeping to machine shop theory: "Adult Education Courses Offered," *Air Scoop*, February 17, 1956.

142 after he made her the offer: Stradling, "Retired Engineer Remembers Segregated Langley."

143 clamber onto the catwalk: Golemba, "Human Computers," 102.

143 published in September 1958: K. R. Czarnecki and Mary W. Jackson, "Effects on Nose Angle 515 and Mach Number on Transition on Cones at Supersonic Speeds," Langley Aeronautical Laboratory, September 1958.

143 suggested that she enroll: Stradling, "Retired Engineer Remembers Segregated Langley."

144 sue the University of Virginia: "Kitty O'Brien Joyner," LAC.

144 only two female engineering graduates in its history: "Woman Engineer Gets Post with RCA Victor Company," *Norfolk Journal and Guide*, November 15, 1952.

144 to enter Hampton High School: Stradling, "Retired Engineer Remembers Segregated Langley."

144 "special permission": Ibid.

145 in the spring of 1956: Jackson Personnel File.

145 dilapidated, musty old building: Stradling, "Retired Engineer Remembers Segregated Langley."

146 In general, the black men at Langley: Thomas Byrdsong, personal interview, October 4, 2014.

146 blue-collar mechanics, model makers, and technicians: Ibid.

147 they escaped to a black-owned restaurant: Williams interview.

CHAPTER 15: YOUNG, GIFTED, AND BLACK

149 her daily job: Christine Darden, personal interview, May 3, 2012.

150 she perused the newspapers: Ibid.

151 "Red-Made Satellite Flashes Across U.S": *Daily Press*, October 5, 1957.

151 "Sphere Tracked in 4 Crossings Over US": *New York Times*, October 5, 1957.

151 "Project Greek Island": "The Secret Bunker Congress Never Used," National Public Radio, March 26, 2011, http://www.npr.org/2011/03/26/134379296/the -secret-bunker-congress-never-used.

151 1992 exposé: Ted Gup, "The Ultimate Congressional Hideway," *Washington Post*, May 31, 1992.

152 "small ball in the air": David S. F. Potree, "One Small Ball in the Air: October 4, 1957–November 3, 1957," *NASA's Origins and the Dawn of the Space Age, Monographs in Aerospace History 10*, National Aeronautics and Space Administration, September 1998.

152 many of them, hundreds perhaps: Only years later would the United States learn that the size and capability of the Soviet arsenal was greatly exaggerated. See McDougall, *The Heavens and the Earth*, 250–53.

153 We can't let them beat us: Darden interview.

153 Radio Moscow announced: "Reds List Sputnik Time for Little Rock," *Washington Post*, October 10, 1957.

154 "I just came to let you all know": Christine Darden, "Growing Up in the South During *Brown v. Board*," *Unbound Magazine*, March 5, 2015.

154 son of a former fire chief: Steven A. Holmes, "Jesse Helms Dies at 86; Conservative Force in Senate," *New York Times*, July 5, 2008.

154 nonexistent science laboratories: Darden, "Growing Up in the South."

154 they would not be good enough: Ibid.

154 "education, honesty, hard work, and character": Wini Warren, *Black Women Scientists in the United States*, 75.

155 Pontiac Hydromatic: *Christine Darden*, (Bloomington: Indiana University Press, 2000), The History Makers, February 26, 2013, http://www.thehistory makers.com/biography/christine-darden.

155 "What did you learn today?": Ibid.

155 priming the carburetor: Ibid.

155 tearing out their stuffing: Ibid.

155 acres of cotton fields: Ibid.

156 released in time for the harvest: Ibid.

156 second-grade student: Ibid.

156 "Julia's parents said she could go": Ibid.

156 one of the best Negro high schools in the country: Rob Neufeld, "Visiting Our Past: The Allen School in Asheville," *Asheville Citizen-Times*, April 27, 2014.

156 Cab Calloway's niece: Ibid.

156 A 1950 graduate named Eunice Waymon: Martha Rose Brown, " 'For Colored Girls': Professor Researching Former School for African-American Female Students," *Times and Democrat*, March 11, 2011.

156 Waves of homesickness: Christine Darden, personal interview, October 10, 2012.

157 Bettye Tillman and JoAnne Smart: "Letters of Intent," *UNCG Magazine*, Spring 2010, http://www.uncg.edu/ure/alumni_magazineT2/2010_spring/feature_lettersofintent.htm.

157 "After careful deliberation": Benjamin Lee Smith, "Report of the Superintendent to the Greensboro City Board of Education regarding *Brown v. Board of Education*," 1956. http://libcdm1.uncg.edu/cdm/ref/collection/CivilRights/id/547.

158 "I've been accepted at Hampton": *Christine Darden*, The History Makers.

158 "Red engineering schools": *Washington Post*, February 23, 1958.

158 Soviet engineering grads were female: Ibid. The article reported that at the same time, women were just 1 percent of American engineering graduates.

159 "civilian army of the Cold War": Sylvia Fries, "The History of Women in NASA, "Women's Equality Day Address, Marshall Space Flight Center, August 23, 1991.

159 took her breath away: *Christine Darden*, The History Makers.

CHAPTER 16: WHAT A DIFFERENCE A DAY MAKES

161 the winking dot of light: *Katherine G. Johnson*, The History Makers, February 6, 2013, http://www.thehistorymakers.com/biography/katherine-g-johnson-42.

161 "One can imagine the consternation": *Reference Papers Relating to a Satellite Study, RA-15032* (Santa Monica, CA: RAND Corp., 1947); F. H. Clauser, *Preliminary Design of a World Circling Spaceship* (Santa Monica, CA: RAND Corp, 1947).

162 a little too far out: Hansen, *Spaceflight Revolution*, 17.

162 "backward peasantry": Roland, *Model Research*, 262.

162 "First in space means first, period": McDougall, *The Heavens and the Earth*.

162 Americans were ahead of the Russians: Ibid., 131.

163 "revolutionary advances for atmospheric aircraft": W. Hewitt Phillips, *A Journey into Space Research: Continuation of a Career at NASA Langley Research Center* (Washington, DC: NASA History Office, 2005), 1.

163 ended by a 1958 NACA headquarters edict: Ibid.

163 "dirty word": Hansen, *Spaceflight Revolution*, 17.

163 hard-pressed to find books on spaceflight: Chris Kraft, *Flight: My Life in Mission Control* (New York: Plume, 2002), 63.

164 an advanced version of the X-15 rocket plane: Hansen, *Spaceflight Revolution*, 356–61.

164 "notoriously freethinking": Hansen, *Spaceflight Revolution*, 197.

165 had come online in 1955: Roger Launius, "NACA-NASA and the National Unitary Wind Tunnel Plan, 1945–1965," 40th AIAA Aerospace Sciences Meeting & Exhibit, January 14–17, 2002, Reno, Nevada, http://crgis.ndc.nasa.gov/crgis/images/d/de/A02-14248.pdf.

165 "nearly every supersonic airplane": Launius, "NACA-NASA and the National Unitary Wind Tunnel Plan."

165 downsized to the new office in 1251: Langley Aeronautical Laboratory Telephone Directory, 1956.

168 gave Miriam Mann's daughter a shiny new penny: Harris interview.

168 Eunice Smith volunteered: "Association Thanks Helpers at Party," *Air Scoop*, January 2, 1953.

168 Dorothy Vaughan's children counted the days: Kenneth Vaughan, personal interview, April 4, 2014.

168 Langley Air Force Base: Mark St. John Erickson, "Colorblind Sword: Military Has Become Model for Race Reform, Experts Say," *Daily Press*, July 28, 1998.

169 "Integration anywhere means destruction everywhere": Donald Lambro, "Pulitzer-winning Journalist Mary Lou Forbes Dies at 83," *Washington Times*, June 29, 2009. Archival footage of Almond's 1958 inaugural speech can be viewed at https://vimeo.com/131577357.

169 "How can Senator Byrd": John B. Henderson, "Henderson Speaks: Closing Schools No Way to Cope with Sputniks," *Norfolk Journal and Guide*, November 23, 1957.

169 forcing each of those Virginia school districts to integrate: Smith, *They Closed Their Schools*, 144.

169 "the 'separate but equal' education of the Negroes marks time": James Rorty, "Virginia's Creeping Desegregation: Force of the Inevitable," *Commentary Magazine*, July 1956. Rorty's article offers a fascinating snapshot of Virginia's struggle with desegregation in the years just after *Brown v. Board of Ed*.

170 "Eighty percent of the world's population is colored": Paul Dembling to file, July 7, 1956.

171 NASA: For years, the folks who had worked for Langley prior to 1958 could be distinguished by the fact that they said the name of the new agency after the fashion of the old one, pronouncing each letter separately: "the N-A-S-A."

171 "to provide for the widest": The Space Act of 1958, http://www.hq.nasa.gov /office/pao/History/spaceact.html.

171 "the bearer of a myth": McDougall, *The Heavens and the Earth*, 376.

171 "the West Area Computers Unit is dissolved": Floyd L. Thompson to All Concerned, "Change in Research Organization," May 5, 1958, NARA Phil.

173 "She was the smartest of *all* the girls": Johnson interview, September 17, 2011.

CHAPTER 17: OUTER SPACE

175 "This is not science fiction": *Introduction to Outer Space: An Explanatory Statement Prepared by the President's Science Advisory Committee*, January 1, 1958. The pamphlet's words "the thrust of curiosity that leads men to try to go where no one has gone before" inspired the well-known introduction to the television series *Star Trek*.

175 "As everyone knows": Ibid.

176 The only real reference: Forest Ray Moulton, *Introduction to Celestial Mechanics* (New York: Macmillan, 1914).

178 "Present your case, build it, sell it so they believe it": Claiborne R. Hicks, interview with Kevin M. Rusnak, JSC, April 11, 2000.

178 months, even years in the making: This long lead time was underscored every time I read through an NACA or NASA research report: the cover lists the date of publication, but the date that the researchers actually completed and submitted the research for review is included at the end of the body of the report.

178 Katherine sat down with the engineers to review: Johnson interview, September 15, 2015.

179 "Why can't I go to the editorial meetings?": Johnson interview, September 27, 2013.

179 "Girls don't go to the meetings": Ibid.

179 "Is there a law against it?" Ibid.

179 laws restricting her ability to apply for a credit card: Diana Pearl, "Rights Women Didn't Used to Have," Marie Claire.com, August 18, 2014, http://www .marieclaire.com/politics/news/a10569/things-women-couldnt-do-1920/.

180 "Women Scientists": This 1959 Langley file photo was simply labeled "Women Scientists." It was published in James Hansen's 1995 book *Spaceflight Revolution*, p. 105, but without names. The NASA Cultural Resources website selected the photo for its July 2013 "Mystery Archive," inviting visitors to help identify the women portrayed; see http://crgis.ndc.nasa.gov/historic/Mystery _Archives_2013.

180 Five out of the six women . . . worked in PARD: Langley Research Center Telephone Directory, 1959, LAC.

180 in 1948, fresh out of Randolph-Macon Women's College: Dorothy B. Lee, interview with Rebecca Wright, JSC, November 10, 1999.

180 he invited her to become a permanent member of his branch: Ibid.

180 authored one report, coauthored seven more: Dorothy B. Lee, "Flight Performance of a 2.8 KS 8100 Cajun Solid-Propellant Rocket Motor," Langley Aeronautical Laboratory, Janaury 21, 1957, NTRS.

180 "Do you believe": Lee interview.

181 "inveterate wind tunnel jockeys": Becker, *The High Speed Frontier*, 19.

181 "can't-hack-it engineers": Gloria R. Champine, personal interview, April 2, 2014.

181 "they were all the same": Johnson interview, December 27, 2010.

182 "Let her go": Johnson interview, September 27, 2013.

CHAPTER 18: WITH ALL DELIBERATE SPEED

183 from erudite and obscure to obvious and spectacular: Yanek Mieczkowski, Eisenhower's Sputnik Moment: The Race for Space and World Prestige (Ithaca: Cornell University Press, 2013), 235.

184 "a dull bunch of gray buildings": Charles Murray and Catherine Bly Cox, *Apollo* (Burkittsville, MD: South Mountain Books, 2004), 322.

184 "So far as the future histories of this state": Lenoir Chambers, "The Year Virginia Closed the Schools," *The Virginian-Pilot*, January 1, 1959. The *Virginian-Pilot* was the only white newspaper in Virginia to take an editorial stand in favor of school desegregation.

184 A total of ten thousand of the shut-out students: Kristen Green, *Something Must Be Done About Prince Edward County: A Family, a Virginia Town, a Civil Rights Battle* (New York: HarperCollins, 2015), 1347–49.

185 salutatorian of Carver High School's class of 1958: "Peninsula Social Whirl," *Norfolk Journal and Guide*, June 14, 1958.

186 drifting toward the social sidelines: Katherine Goble Moore, personal interview, February 7, 2015.

186 Mary Jackson had been one of his student teachers: James A. Johnson, personal interview, June 11, 2011.

186 "Ladies, he's single": Johnson interview, March 13, 2011.

187 arriving together at church: James A. Johnson.

189 "bootlegged": Cox and Murray, *Apollo*, ch.1

189 "computing runs": Ibid.

189 "a hell of a lot of fun": Ibid.

189 thousands of woman-hours computing ballistics trajectory tables: LeAnn Erickson's documentary Top Secret Rosies provides a detailed look at the University of Pennsylvania. See https://www.facebook.com/topsecretrosies.

190 "Let me do it": *Katherine Johnson*, The History Makers.

192 "Katherine should finish the report": Warren, *Black Women Scientists in the United States*, 143.

192 by a female author: Ted Skopinski and Katherine G. Johnson, "Determination of Azimuth Angle at Burnout for Placing a Satellite over a Selected Earth Position," Langley Research Center, 1960.

CHAPTER 19: MODEL BEHAVIOR

193 The bar had been set the year before: "Congratulations . . ." *Air Scoop*, July 1, 1960.

193 "The car and driver together": *Soapbox Derby Rules* 1960.

194 Levi and his competitors: "Hampton Youth Captures Area Derby Championship," *Norfolk Journal and Guide*, July 2, 1960.

194 nine-hundred-foot racecourse: Ibid.

194 girls weren't allowed to race: Paul Dickson, "The Soap Box Derby," *Smithsonian Magazine*, May 1995. Girls did not compete in the derby until the 1970s.

194 one of fifty thousand boys: "Derby Day Is Your Day!" *Boy's Life*, February 1960, 12.

195 "A Study of Air Flow in Scaled Dimensions": "Science Fair Held at Y. H. Thomas Jr. High," *Norfolk Journal and Guide*, March 31, 1962.

195 "Soapbox what?": Janice Johnson interview.

197 Emma Jean was valedictorian: Golemba, "Human Computers," 39.

197 produced several research reports: "Report Listing from December 1949–October 1981," Unitary and Continuous-Flow Hypersonic Tunnels, LAC, http://crgis.ndc.nasa.gov/crgis/images/a/aa/1251-001.pdf.

197 National Council of Negro Women: "Girls' Group Hears Talk by 2 Women Engineers," *Norfolk Journal and Guide*, February 16, 1963.

197 "The Aspects of Engineering for Women": Ibid.

198 one of the largest minority troops on the peninsula: "Girl Scout Pioneers Honored During Tribute in Hampton," *Norfolk Journal and Guide*, November 6, 1985.

198 troop leader: Janice Johnson interview.

198 So Mary enlisted the help of Helen Mulcahy: Ibid.

198 take Janice trekking: Ibid.

199 an enthusiastic crowd of four thousand: "Hampton Youth Captures Area Derby Championship."

199 clear, warm, just enough of a breeze: Newport News, Virginia, Historical Weather, Almanac.com, July 3, 1962.

199 Officials weighed and inspected each car: "Hampton Youth Captures Area Derby Championship."

199 "a drop of oil on each wheel bearing": Ibid.

199 seventeen miles per hour: Ibid.

200 the slimness of his machine: Ibid.

200 "I want to be an engineer like my mother": "Hampton Youth Captures Area Derby Championship."

200 a spot at the national All-American Soap box Derby: Ibid.

200 in front of seventy-five thousand fans: "Derby Day Is Your Day!"

200 "first colored boy in history": "Hampton Youth Captures Area Derby Championship."

200 the donations started rolling in: "Citizens Honor Local Soap Box Derby Champ," *Norfolk Journal and Guide*, August 27, 1960.

CHAPTER 20: DEGREES OF FREEDOM

201 reliability tests on the Mercury capsule: Loyd S. Swenson Jr., James M. Grimwood, and Charles C. Alexander, *This New Ocean: A History of Project Mercury* (Washington, DC: NASA, 1989), 256.

201 three hundred had joined the demonstration: "The Greensboro Sit-In," History.com, http://www.history.com/topics/black-history/the-greensboro-sit-in

201 "Dear Mom and Dad": John "Rover" Jordan, "This Is Portsmouth," *Norfolk Journal and Guide*, June 8, 1963.

202 offering her a job as a hostess: Dr. William R. Harvey, "Hampton University and Mrs. Rosa Parks," *Daily Press*, February 23, 2013.

202 seven hundred: Arriana McLymore, "A Silenced History; Hampton's Legacy of Student Protests," *Hampton Script*, November 6, 2015.

202 until the owners shut down their establishments: "Hampton 'Sit-down': Students Seek Service; 5 & 10 Counter Closes," *Norfolk Journal and Guide*, February 20, 1960.

202 five hundred students staged a peaceful protest: Jimmy Knight, "Hamptonians Vow: Jail Will Not Stop Student Protests," *Norfolk Journal and Guide*, March 5, 1960.

202 "We want to be treated as American citizens": Ibid.

202 walking door-to-door in black neighborhoods: Christine Darden, personal interview, April 30, 2012.

203 alive, breathless even: Hammond interview.

203 the astronauts were contributing to the students' organizing activities: Ibid. Though I could never find documents to support this, many in Ann Vaughan Hammond's circle of friends had heard the rumor; her memory of the rumor and of the enthusiasm it engendered among the students was vivid.

203 reopening Norfolk, Charlottesville, and Front Royal schools: Lenoir Chambers, "The Year Virginia Opened the Schools," *Virginian-Pilot*, December 31, 1959. Chambers, the editor-in-chief of the *Virginian-Pilot*, was awarded the Pulitzer Prize for this and the eleven other editorials he wrote over the course of 1959.

204 "The only places on earth known not to provide free public education": Smith, *They Closed Their Schools*, 190. Smith's book is perhaps the best historical account of the Prince Edward County school situation.

204 Marjorie Peddrew and Isabelle Mann: Thompson, "Change in Research Organization."

208 aerodynamic, structural, materials, and component tests: Hansen, *Spaceflight Revolution*, 60.

208 "We could have beaten them": Kraft, *Flight*, 132. See also Robert Gilruth, interview with David DeVorkin and John Mauer, National Air and Space Museum, March 2, 1987, part 6, http://airandspace.si.edu/research/projects/oral-histories/TRANSCPT/GILRUTH6.HTM.

208 1.2 million tests, simulations, investigations: "Webb Receives Safety Award," *Air Scoop*, June 30, 1961.

208 Have the chimpanzee: Swenson, Grimwood, and Alexander, *This New Ocean*, 317.

208 Forty-five million Americans: Ibid.

209 116.5 miles above Earth: Ibid., 355.

209 fifteen minutes and twenty-two seconds and covered 303 miles: Ibid.

209 "I believe that this nation should commit itself": John F. Kennedy, "Urgent National Needs: A Special Message to Congress by President Kennedy," May 25, 1961, http://www.presidency.ucsb.edu/ws/?pid=8151.

209 required a team of eighteen thousand people: Swenson, Grimwood, and Alexander, *This New Ocean*, 508.

210 Twenty locations made the short list: James Grimwood, Project Mercury: A Chronology (Washington D.C.: National Aeronautics and Space Administration, 1963), 147.

210 "was going to be badly understaffed": Harold Beck, "The History of Mission Planning for Manned Spaceflight," unpublished document in author's possession.

210 asked to transfer to Houston: Katherine Johnson, personal interview, September 27, 2013.

210 "five qualified young women": Beck, "History."

CHAPTER 21: OUT OF THE PAST, THE FUTURE

213 ninety-five-foot-high, 3.5-million-horsepower: Colin Burgess, *Friendship 7: The Epic Orbital Flight of John H. Glenn Jr.* (New York: Springer Praxis Books, 2015).

214 "a Rube Goldberg device on top of a plumber's nightmare": Swenson, Grimwood, and Alexander, *This New Ocean*, 411.

214 running miles each day: Tom Wolfe, *The Right Stuff*, 128.

214 water egress from the capsule: "Astronaut Training at Langley," http://crgis.ndc.nasa.gov/historic/Astronaut_Training.

214 hundreds of simulated missions: Kraft, *Flight*.

215 conspired to push the date: Swenson, Grimwood, and Alexander, *This New Ocean*, 273–83.

216 resisted the computers: David A. Mindell, *Digital Apollo* (Cambridge, MA: The MIT Press, 2008), 175.

217 a black scientist named Dudley McConnell: Sylvia Doughty Fries, *NASA Engineers in the Age of Apollo* (Washington, DC: NASA, 1992).

218 staffed on Project Centaur: Annie Easley, interview with Sandra Johnson, JSC, August 21, 2001.

218 a Howard University graduate named Melba Roy: Alice Dunnigan, "Two Women Help Chart Way for the Astronauts," *Norfolk Journal and Guide*, July 6, 1963.

218 Hoover had worked at the Weather Bureau: Ibid.

219 calculations that were used in Project Scout: Golemba, "Human Computers," 121.

219 "mad scientists": Hansen, *Spaceflight Revolution*, 345.

221 a blazing 1 kilobyte per second: Saul Gass, "Project Mercury Real-Time Computational and Data-Flow System," IBM, 1961.

222 proposed blaming it on the Cubans: James Bamford, *Body of Secrets* (New York: Anchor Books, 2001) Kindle ed., loc. 1525.

223 phone-book-thick stacks of data sheets: Johnson interview, September 27, 2013.

223 three-orbit mission: Swenson, Grimwood, and Alexander, *This New Ocean*.

223 It took a day and a half: Johnson interview, September 27, 2013.

223 February 20 dawned with clearing skies: Burgess, *Friendship 7*.

223 One hundred thirty-five million people: Ibid.

224 3,000-degree Fahrenheit temperatures: Ibid.

224 off by forty miles: Ibid.

224 "our Ace of Space": Izzy Rowe, "Izzy Rowe's Notebook," *Pittsburgh Courier*, March 10, 1962.

224 Thirty thousand local residents: Hansen, *Spaceflight Revolution*, 77.

225 fifty-car parade: Ibid.

225 traced a twenty-two-mile route: Ibid.

225 Joylette and Dorothy Vaughan's son, Kenneth: Interviews with Joylette Goble Hylick, Kenneth Vaughan, and Christine Mann Darden.

225 sign reading SPACETOWN, USA: Hansen, *Spaceflight Revolution*, 80.

225 "Katherine Johnson: mother, wife, career woman!": "Lady Mathematician Played a Key Role in Glenn Space Flight," *Pittsburgh Courier*, March 10, 1962.

225 "Why No Negro Astronauts?" *Pittsburgh Courier*, March 10, 1962.

CHAPTER 22: AMERICA IS FOR EVERYBODY

227 "America Is for Everybody": US Department of Labor, April 1963.

227 landed on Katherine Johnson's desk in May 1963: John P. Scheldrup to Edward Maher, May 15, 1963, NARA Phil.

227 "occupied positions of responsibility": Ibid.

227 "analyzing lunar trajectories": "America Is for Everybody."

228 in English North America in 1619: Robert Brauchle, "Virginia Changing Marker Denoting Where First Africans Arrived in 1619," *Daily Press*, August 19, 2015. For years, Jamestown was thought to be the first landing place for the "twenty and odd" Africans who were brought as slaves to English-speaking North America, but recent research has revealed that they disembarked at Old Point Comfort in Hampton, site of modern-day Fort Monroe.

228 in 1963 with a twenty-two-orbit flight: Swenson, Grimwood, and Alexander, *This New Ocean*, 494.

228 Dorothy Height, John Lewis, Daisy Bates, and Roy Wilkins: Though the women played a critical behind-the-scenes role helping to organize the day's events, none of them were given a prominent speaking role that day.

228 three hundred thousand people: Branch, *Parting the Waters*, 878.

228 "He's Got the Whole World in His Hands": Marian Anderson onstage at the March on Washington, 1963, https://www.youtube.com/watch?v=2HfNovwcaX8.

228 W. E. B. Du Bois had died early that morning: Branch, *Parting the Waters*, 878.

229 "Dear Mrs. Vaughan": Floyd L. Thompson to Dorothy J. Vaughan, July 8, 1963, Vaughan Personnel File.

229 a gold-and-enamel lapel pin: Ibid.

230 "very few Negroes": Floyd L. Thompson to James E. Webb, December 29, 1961, NARA Phil.

230 "social and economic mobility": Fries, *NASA Engineers in the Age of Apollo*, loc. 1385.

230 with "dreams of working at NASA": Ibid.

234 fell asleep at the wheel: Warren, *Black Women Scientists in the United States*, 144.

CHAPTER 23: TO BOLDLY GO

235 a hundred or so black women: Johnson interview, September 27, 2013.

236 groups of them perching in front of the screen: Ibid.

236 a total of six hundred million people: Scott Christianson, "How NASA's Flight Plan Described the Apollo 11 Moon Landing," Smithsonian.com, November 24, 2015, http://www.smithsonianmag.com/us-history/apollo-11-flight -plan-180957225/?no-ist.

236 four hundred thousand: "NASA Langley Research Center's Contributions to the Apollo Program," n.d., http://www.nasa.gov/centers/langley/news/fact sheets/Apollo.html.

237 the weekend leadership conference: "Alpha Kappa Alpha's 39th Mid-Western Regional Conference at LU," *Langston University Gazette*, July 1969.

237 96 degrees in Hampton: Weather History for Hampton, Virginia, Farmer's Almanac (accessed via Almanac.com).

237 a car full of sorority members: Johnson interview, September 15, 2015.

237 pink-and-green-clad women: Pink and green are the official colors of Alpha Kappa Alpha.

237 the most promising young women: "Alpha Kappa Alpha's 39th Mid-Western Regional Conference at LU," *Langston University Gazette*, July 1969.

237 a full-time job training center: Ibid.

237 Hillside's thirty-three rooms: Matt Birkbeck, *Deconstructing Sammy: Music, Money and Madness* (New York: HarperCollins, 2008), 162.

238 bought the land with his Jewish business partner: Wendy Beech, *Against All Odds: Ten Entrepreneurs Who Followed Their Hearts and Found Success* (New York: Wiley, 2002), 204.

238 The Hillside advertised: The Hillside was a mainstay of these black publications, its small black-and-white ad appearing regularly: "Pennsylvania's Famous Resort Hotel HILLSIDE INN in the Heart of the Poconos Mountains Air Conditioned Rooms, Swimming Pool. Color TV . . ."

238 sweet potato pie and peach cobbler for dessert: Lawrence Louis Squeri, *Better in the Poconos: The Story of Pennsylvania's Vacationland* (University Park, PA: Pennsylvania State Press, 2002), 182.

238 students at black colleges in the South: Ibid.

239 set the candle on fire at 9:37 a.m.: CBS News coverage of the launch of Apollo 11, July 17, 1969, https://www.youtube.com/watch?v=yDhcYhrCPmc.

239 Walter Cronkite to wield the jargon: Ibid.

240 the mighty Saturn V rocket consumed: Ibid.

249 $24 billion: Ibid.

241 perceived mistreatment of Ed Dwight: Richard Paul and Steven Moss, *We Could Not Fail: The First African Americans in the Space Program* (Austin: University of Texas Press, 2015), loc. 1902.

243 a celebrity NAACP civil rights fund-raiser: Nichelle Nichols interview with Neil deGrasse Tyson, StarTalk Radio, July 11, 2011, http://startalkradio.net/show/a-conversation-with-nichelle-nichols/.

243 "her greatest fan": Ibid.

243 face-to-face with Dr. Martin Luther King Jr.: Ibid.

243 fourth in command of the ship: Ibid.

243 asked him to tear up the resignation letter: Ibid.

244 curiosity always bested fear: Moore interview.

244 Then, finally, at 10:38 p.m.: CBS News coverage of Apollo 11 lunar landing," https://www.youtube.com/watch?v=E96EPhqT-ds.

244 Neil Armstrong handicapped the odds: Neil Armstrong, interview with Alex Malley, 2011, https://www.youtube.com/watch?v=jfj2jqpst_Q.

245 "You have to expect progress to be made": Johnson interview, December 27, 2010.

245 born at a time when the odds were more likely: 1920 US Census, Statistic of the Population.

245 circling the Moon every fifty-nine minutes: Richard Orloff, *Apollo by The Numbers: A Statistical Reference* (Washington, DC: National Aeronautics and Space Administration, 2005), http://histry.nasa.gov/SP-4029/Apollo_18-01_General_Background.htm.

246 plot a course to Mars: Johnson interview, January 3, 2011; Harold A. Hamer and Katherine G. Johnson, "Simplified Interplanetary Guidance Procedures Using Onboard Optical Measurements," Langley Research Center, May 1972, NTRS.

246 "grand tour" of the outer planets: J. W. Young and M. E. Hannah, "Alternate Multiple-Outer-Planet Missions Using a Saturn-Jupiter Flyby Sequence,"

Langley Research Center, December 1973, NTRS. Marge Hannah and John Young received NASA achievement awards for their work on this paper. See: "Reid Award Committee Selects Best Directorate Papers for Honorable Mentions," *Langley Researcher*, November, 1974, 5; John Worth Young Obituary, http://www.memorialsolutions.com/sitemaker/memsol_data/2061 /1292572/1292572_2061.pdf.

EPILOGUE

248 "I loved every single day of it": Johnson interview, December 27, 2010.

248 her greatest contribution to the space program: Johnson interview, September 27, 2013.

248 So what do you do when the computers go out?: Warren, *Black Women Scientists in the United States*, 144.

248 the first of a series of reports: Harold A. Hamer and Katherine G. Johnson, "An Approach Guidance Method Using a Single Onboard Optical Measurement," NASA Langley Research Center, October 1970.

249 with Earth's terminator: Nancy Atkinson, "13 Things That Saved Apollo 13, Part 6: Navigating by Earth's Terminator," UniverseToday.com, April 16, 2010.

249 "They are loud in their praise": James L. Hicks, "Negroes in Key Roles in US Race for Space: Four Tan Yanks on Firing Team," *New York Amsterdam News*, February 8, 1958.

250 A STEM institute bearing her name: The Alpha Academy in Fayetteville, North Carolina, plans to unveil its Katherine G. Johnson STEM Institute in 2016.

251 "Rockets, moon shots, spend it on the have-nots": James Nyx Jr. and Marvin Gaye, "Inner City Blues," *What's Going On*, New York: Sony/ATV Music Publishing, 1971.

252 "pollution, ecological damage, energy shortages, and the arms race": Robert Ferguson, *NASA's First A: Aeronautics from 1958 to 2008* (Washington, DC: National Aeronautics and Space Administration, 2012).

252 "salt on the wounds": Ibid.

252 "big fat money pot": Alan Wasser, "LBJ's Space Race: What We Didn't Know Then, Part Two," The Space Settlement Institute, June 27, 2005, http://www .thespacereview.com/article/401/1.

252 cancel its supersonic transport program: Christine M. Darden, "Affordable Supersonic Transport: Is It Near?" Japan Society for Aeronautical and Space Sciences lecture, Yokohama, Japan, October 9–11, 2002.

252 an "Apollo moment": Hansen, *Spaceflight Revolution*, 102.

252 "setting dogs to barking": Lawrence R. Benson, *Quieting the Boom: The Shaped Sonic Boom Demonstrator and the Quest for Quiet Supersonic Flight* (Washington, DC: National Aeronautics and Space Administration, 2013), 8.

252 "death of pets and the insanity of livestock": Ibid, 7.

253 "164 million": "Exploring in Aeronautics: An Introduction to Aeronautical Sciences Developed at the NASA Lewis Research Center," NASA Lewis Research Center, 1971, 1.

253 Langley announced a sweeping reorganization: Edgar M. Cortright, "Reorganization of Langley Research Center," September 24, 1970.

253 to a total of 3,853 from its peak of 4,485: Hansen, *Spaceflight Revolution*, 102.

253 "routine, quick-reaction and economical access to space": "Tenth Anniversary of John Glenn's Space Flight Observed," *Langley Researcher*, March 3, 1972.

254 Mary took FORTRAN classes: Jackson Personnel File.

254 She made so many speeches: "Speaker's Bureau," *Langley Researcher*, February 20, 1976.

254 "We have to do something like this": "Personnel Profiles," *Langley Researcher*, April 2, 1976.

255 organized the retirement party for Kazimierz Czarnecki: "Retirement Parties," *Langley Researcher*, December 15, 1978.

255 papers to her name: Mary Jackson, "Mary W. Jackson, Federal Women's Program Coordinator," LHA, October 1979.

255 This was a contrast with Goddard: Dunnigan, "Two Women Chart Way for Astronauts."

255 "to place a woman in at least one:" Edgar Cortright to Grove Webster, "NASA Plans to Attract More Qualified Women to Government Positions," June 11, 1971, NARA Phil.

255 restricted women to playing during the workday: Sharon H. Stack, personal interview, April 22, 2014.

255 she had probably hit the glass ceiling: Champine interview.

256 instrumental in bringing the separate: Mary Winston Jackson Obituary program, February 17, 2005, in author's possession.

256 equal opportunity employment counselor: "Meet Your EEO Counselors: Mary Jackson," *Langley Researcher*, June 23, 1972.

256 Langley's Federal Women's Program Advisory Committee: "Advisory Committee," *Langley Researcher*, May 11, 1973.

257 "fantasy that men were uniquely gifted": Fries, "The History of Women in NASA."

258 "everybody's daddy had a plane": Gloria Champine, personal interview, July 23, 2014.

258 the "crazy things": Gloria Champine, "XB-15: First of the Big Bombers of World War II," NASA History website, http://crgis.ndc.nasa.gov/historic /XB-15. Gloria's father's crew worked with the NACA's chief test pilot, Melvin Gough, and a young Robert Gilruth to produce the report "Stalling Characteristics of the Boeing XB-15 Airplane (Air Corps No. 35-277), by M. N. Gough and R. R. Gilruth.

258 "They kept testing you": Champine interview.

258 "hard head and strong shoulders and back": Gloria Champine, interview with Sandra Johnson, JSC, May 1, 2008.

259 Gloria marched her over to meet: "EEO Highlights," *Langley Researcher*, July 20, 1973.

259 "stay away from the woman stuff": Champine interview, May 1, 2008.

259 It was a decision that helped her: Claudia Goldin, "The Female Labor Force and American Economic Growth, 1890–1980," in Stanley L. Engerman and Robert E. Gallman, eds., *Long-Term Factors in American Economic Growth* (Chicago: University of Chicago Press, 1986), 557–604.

260 "We always thought it was so cool": Wanda Jackson, telephone interview, February 15, 2016.

260 "The Peninsula recently lost a woman of courage": Gloria Champine, "Mary Jackson," NASA website, February 2005, http://crgis.ndc.nasa.gov/crgis /images/4/4a/MaryJackson.pdf.

261 It was "deadly": Fries, *NASA Engineers in the Age of Apollo*, loc. 1741.

261 knocked off the board by a black man: *Christine Darden*, The History Makers.

261 "Why is it that men get placed into engineering groups": Darden interview.

261 "Well, nobody's ever complained": Ibid.

261 had been an "excellent mathematician": John Becker, personal interview, August 10, 2014; Golemba, "Human Computers," 4.

262 self-described "wing man": "David Earl Fetterman Jr.," *Daily Press*, March 5, 2003.

262 It took three years of work: Christine M. Darden, "Minimization of Sonic-Boom Parameters in Real and Isothermal Atmospheres," Langley Research Center, 1975.

262 sixty technical publications and presentations: Warren, *Black Women Scientists in the United States*, 78.

262 seven men and one woman: Christine Darden, personal interview, February 12, 2012; Christine Darden, "Growing Up in the South During Brown v. Board," Old Dominion University Commencement Address, December 15, 2012, http://justiceunbound.org/carousel/growing-up-in-the-south-during-brown-v-board/.

262 "juggling the duties of Girl Scout mom": Warren, *Black Women Scientists in the United States*, 77.

263 Gloria Champine admired Christine Darden's intelligence: Gloria Champine, personal interview, July 23, 2014.

263 "It involved a promotion": Hammond interview, April 4, 2014.

264 was given to Roger Butler: Cortright, "Reorganization of the Langley Research Center."

264 Sara Bullock, the East Computer: Ibid.

264 In 1971, there were still no female: Ibid.

264 Only reluctantly did she agree: Hammond interview, April 3, 2014.

BIBLIOGRAPHY

SOURCES

Archival Sources

Daily Press Archives, Newport News Library, Main Street Branch, Newport News, Virginia. Available as microfiche only.

Farmville Herald Archives, Longwood College, Farmville, Virginia. Available as microfiche only.

Hampton University Archives, Hampton, Virginia.

Langley Research Center Archives, Hampton, Virginia.

National Aeronautics and Space Administration History Office, Washington, DC (NASA HQ). http://history.nasa.gov/hqinventory.pdf.

National Archives and Records Administration (NARA), Regional Facilities:

Philadelphia, Pennsylvania: Records of the National Aeronautics and Space Administration (RG 255); Records of the US Civil Service Commission (RG 146); Records of the War Manpower Commission (RG 211).

College Park, Maryland: Records of the National Aeronautics and Space Administration; Records of the US Department of Education; Records of the Fair Employment Practices Commission.

Fort Worth, Texas: Records of the National Aeronautics and Space Administration (RG 255), Project Mercury Working Paper Series, nos. 104, 106, 191, 207, 212, and 217.

St. Louis, Missouri: National Personnel Records Center (NPRC). NPRC documents for deceased Civil Service employees available upon written request. All personnel records cited in the text came from this archive.

West Virginia State University Archives, Institute, West Virginia.

Online Sources

Ancestry.com. Ancestry.com was the source for Census Bureau data; marriage, birth, and death records; and local telephone directories.

Baltimore Afro-American. Archive accessed through Google Books.

"Hampton Roads Embarkation Series, 1942–1946," US Army Signal Corps Photograph Collection, Library of Virginia (HRE), http://www.lva.virginia.gov /exhibits/treasures/arts/art-m12.htm.

The History Makers. This searchable video archive is dedicated to the oral histories of prominent contemporary African Americans. Interviews consulted for the book include Christine Darden, Katherine Johnson, Woodrow Whitlow, and James E. West; http://www.thehistorymakers.com/taxonomy/term/7298.

Johnson Space Center Oral History Project (JSC), http://www.jsc.nasa.gov/history /oral_histories/oral_histories.htm. Oral histories consulted in this collection include Harold Beck, John Becker, Jerry Bostick, Stefan Cavallo, Gloria Champine, Beverly Swanson Cothren, Annie Easley, John H. Glenn, Jane Hess, Claiborne Hicks, Shirley H. Hinson, Eleanor Jaehnig, Harriet Jenkins, Eldon Kordes, Christopher Kraft, Mary Ann Johnson, Dorothy B. Lee, Glynn Lunney, Charles Matthews, Catherine T. Osgood, Emil Schiesser, Alan Shepard, Milton Silveira, and Ruth Hoover Smull.

NASA History Series publications (NH). NASA's lineup of history publications is nothing short of spectacular. Most, including each publication in the list of books below, are available for free in PDF and ebook formats at http://history .nasa.gov/series95.html.

NASA Langley Archives Collection (LAC), http://crgis.ndc.nasa.gov/historic /Langley_Archives_Collection. The following resources were consulted: Langley Employee Newsletters (*LMAL Bulletin* (1942–1944); *Air Scoop* (1945–1962); *Langley Researcher* (1963–present)); Langley telephone directories; Oral Histories and Interviews (oral histories and interviews consulted for the book include Ira Abbott, John Becker, Sherwood Butler, T. Melvin Butler, Mary Jackson, W. Kemble Johnson, Arthur Kantrowitz, and Pearl Young); P-51 Mustang Archives Collection; and Langley Historic Site and Building pages.

NASA Langley Youtube channel. Videos consulted on this channel include interviews with Christine Darden, W. Hewitt Philipps, Richard Whitcomb, and group interviews with former computers (*When Computers Were Human* and *Panel Discussion with Women Computers,* moderated by James R. Hansen).

NASA Technical Reports Server (NTRS), http://ntrs.nasa.gov/. This fully search-able database contains most of the research reports produced by the NACA and NASA from their inception to the present day.

National Visionary Leadership Project (NVLN). This video archive houses inter-views with prominent African Americans over the age of seventy. Interviews consulted include Oliver Hill and Katherine Johnson.

New York Age. Archives accessed through Newspapers.com.

Norfolk Journal and Guide. Archives accessed through the Library of Virginia website, http://www.lva.virginia.gov/.

Pittsburgh Courier. Archives accessed through Newspapers.com.

PERSONAL INTERVIEWS

John Becker, Lynchburg, VA; August 20, 2014.

George M. Brooks, Newport News, VA; July 13, 2014.

Thomas Byrdsong, Newport News, VA; October 4, 2014.

Gloria R. Champine, Newport News, VA; January 24, 2014; April 2, 2014.

Robert S. Conte, White Sulphur Springs, WV; September 12, 2013.

Christine M. Darden, Hampton, VA; May 3, 2012.

Joylette Hylick Goble, Mount Laurel, NJ; October 10, 2011.

Ann Vaughan Hammond, Hampton, VA; April 2, 2014; June 30, 2014.

Miriam Mann Harris, Winston-Salem, NC.

Jane Hess, Newport News, VA.

Wythe Holt, Hampton, VA: July 20, 2014.

Wanda Jackson, Hampton, VA; February 15, 2016.

Eleanor Jaehnig, Hampton, VA; March 7, 2014.

James A. Johnson, Newport News, VA; June 11, 2011.

Janice Johnson, Hampton, VA; April 3, 2014.

Katherine G. Johnson, Newport News, VA; December 27, 2010; March 6, 2011; March 11, 2011; September 17, 2011; September 27, 2011; September 27, 2013.

Edwin Kilgore, Newport News, VA; April 3, 2014.

Kathaleen Land, Hampton, VA; December 19, 2010.

Janet Mackenzie, October 9, 2015, Newport News, VA.

Katherine Goble Moore, Greensboro, NC: April 13, 2014; July 7, 2014; February 7, 2015.

Christine Richie, Newport News, VA; July 20, 2014.

Debbie Schwarz Simpson, September 12, 2012.

Sharon Stack, Gloucester, VA; April 22, 2014.

Elizabeth Kittrell Taylor, Yorktown, VA; July 12, 2014.

Donna Speller Turner, March 8, 2014.

Kenneth Vaughan, Hampton, VA; April 2, 2014.

Leonard Vaughan, Hampton, VA; April 23, 2014.

Michelle Webb, Hampton, VA; February 19, 2016.

Barbara Weigel, Newport News, VA; April 2, 2014.

Jerry Woodfill, Houston, TX; April 29, 2016.

ORAL HISTORY TRANSCRIPTS

Unpublished Documents

Beck, Harold. "Project Mercury Planning Activities from 1958 through 1962," May 2016. Unpublished document in author's possession.

Beck, Harold. "Organization Timeline," May 2016.

Champine, Gloria. *He's Got the Right Stuff*, 2014.

Fox, Dewey W. *A Brief Sketch of the Life of Miss Dorothy L. Johnson*. West Virginia African Methodist Episcopal Sunday School Convention, 1926. Pamphlet in author's possession.

Jackson, Mary. Obituary.

Newsome Park Reunion, September 12, 2005.

Newsome Park Reunion: The Legacy of a Village, September 6, 2006.

"Notes on Space Technology," Langley Research Center, 1958, NTRS, http://ntrs .nasa.gov/archive/nasa/casi.ntrs.nasa.gov/19740074640.pdf.

Vaughan, Dorothy. Biography, undated.

SELECTED BOOKS AND MONOGRAPHS

Anderson, Jervis. *A. Philip Randolph: A Biographical Portrait*. Berkeley: University of California Press, 1986.

Anderson, Karen. *Wartime Women: Sex Roles, Family Relations, and the Status of Women During World War II*. Westport, CT: Greenwood Press, 1981.

Baals, Donald D. and William R. Corliss. *Wind Tunnels of NASA*. Washington, DC: NASA History Office, 1981.

Becker, John V. *The High Speed Frontier: Case Histories of Four NACA Programs, 1920–1950*. Washington, DC: National Aeronautics and Space Administration, 1980.

Bilstein, Roger. *Orders of Magnitude: A History of the NACA and NASA, 1915–1990*. Washington, DC: National Aeronautics and Space Administration, 1989.

Blood, Kathryn. *Negro Women War Workers*. Washington, DC: US Department of Labor, 1945.

Branch, Taylor. *Parting the Waters: America in the King Years 1954–63*. New York: Simon & Schuster, 2007.

Burgess, Colin. *Friendship 7: The Epic Orbital Flight of John H. Glenn, Jr.* New York: Springer Praxis Books, 2015.

Carpenter, M. Scott, Cooper, Gordon L. et al. *We Seven by the Astronauts Themselves*. New York: Simon & Schuster, 1962.

Chambers, Joseph R. *The Cave of the Winds: The Remarkable Story of the Langley Full-Scale Wind Tunnel*. Washington, DC: National Aeronautics and Space Administration, 2014.

Clauser, F. H. *Preliminary Design of a World Circling Spaceship*. Santa Monica, CA: RAND Corp., 1947.

Conte, Robert S. *The History of the Greenbrier: America's Resort*. Parkersburg, PA: Trans Allegheny Books, 1989.

Cooper, Henry S. F. Jr. *Thirteen: The Apollo Flight That Failed*. Baltimore: Johns Hopkins University Press, 1995.

Davis, Thulani. *1959*. New York: Grove Press, 1992.

Deighton, Len. *Goodbye Mickey Mouse*. New York: Knopf, 1982.

Dudziak, Mary L. *Cold War Civil Rights: Race and the Image of American Democracy*. Princeton, NJ: Princeton University Press, 2007.

Engs, Robert Francis. *Freedom's First Generation: Black Hampton, Virginia, 1861–1890*. Philadelphia: University of Pennsylvania, 1979.

Fairfax, Colita Nichols. *Hampton, Virginia*. Charleston, SC: Arcadia Publishing, 2005.

Fries, Sylvia Doughty. *NASA Engineers in the Age of Apollo*. Washington, DC: National Aeronautics and Space Administration, 1992.

Gass, Saul I. "Project Mercury Real-Time Computation and Data Flow System." Washington DC: International Business Machines Corporation, 1961. https://www.computer.org/csdl/proceedings/afips/1961/5059/00/50590033.pdf.

Golemba, Beverly E. "Human Computers: The Women in Aeronautical Research." PhD dissertation, St. Leo College, 1994, available at NASA Cultural Resources, http://crgis.ndc.nasa.gov/crgis/images/c/c7/Golemba.pdf.

Grier, David Alan. *When Computers Were Human*. Princeton, NJ: Princeton University Press, 2005.

Grimwood, J. M., C. C. Alexander, and L. S. Swenson Jr. *This New Ocean: A History of Project Mercury*. Washington DC: National Aeronautics and Space Administration, 1966.

Hansen, James R. *Engineer in Charge: A History of the Langley Aeronautical Laboratory, 1917–1958*. Washington, DC: National Aeronautics and Space Administration, 1987.

Hansen, James R. *Spaceflight Revolution: NASA Langley Research Center from Sputnik to Apollo*. Washington, DC: National Aeronautics and Space Administration, 1995.

Harris, Ruth Bates. *Harlem Princess: The Story of Harry Delanay's Daughter*. New York: Vantage Press, 1991.

Herman, Arthur. *Freedom's Forge: How American Business Produced Victory in World War II*. New York: Random House Publishing Group, 2012.

Holt, Natalia. *Rise of the Rocket Girls: The Women Who Propelled Us, from Missiles to the Moon to Mars*. New York: Little, Brown, 2016.

Hoover, Dorothy. *A Layman Looks With Love At Her Church*. Philadelphia: Dorrance, 1970.

Kalme, Albert P. "Racial Desegregation and Integration in American Education: The Case History of West Virginia State College, 1891–1973." PhD dissertation, University of Ottawa, 1976.

Kessler, James H. et al. *Distinguished African American Scientists of the 20th Century*. Westport, CT: Greenwood Publishing, 1996.

Kraft, Christopher C. *Flight: My Life in Mission Control*. New York: Dutton, 2001.

Kranz, Gene. *Failure Is Not an Option*. New York: Simon & Schuster, 2001.

Krislov, Samuel. *The Negro in Federal Employment: The Quest for Equal Opportunity*. New Orleans: Quid Pro Quo Books, 2012.

Lewis, Earl. *In Their Own Interests: Race, Class and Power in Twentieth-Century Norfolk, Virginia*. Berkeley: University of California Press, 1991.

Margo, Robert. *Race and Schooling in the South, 1880–1950: An Economic History*. Chicago: University of Chicago Press, 1950.

Marsh, Charles F., ed. *The Hampton Roads Communities in World War II*. Chapel Hill: University of North Carolina Press, 1951/2011.

McDougall, Walter A. *The Heavens and the Earth: A Political History of the Space Age*. Baltimore: Johns Hopkins University Press, 1997.

McNeil, Genna Rae. *Groundwork: Charles Hamilton Houston and the Struggle for Civil Rights*. Philadelphia: University of Pennysylvania Press, 1983.

Michener, James A. *Space: A Novel*. New York: Random House, 1982.

Moulton, Forest Ray. *Celestial Mechanics*. New York: The Macmillan Company, 1914.

Muse, Benjamin. *Virginia's Massive Resistance*. Bloomington: Indiana University Press, 1956.

Myrdal, Gunnar. *An American Dilemma: The Negro Problem and Modern Democracy*. New York: Harper, 1944.

Pearcy, Arthur. *Flying the Frontiers: NACA and NASA Experimental Aircraft*. Annapolis, MD: Naval Institute Press, 1993.

Phillipps, William Hewitt. *Journey in Aeronautical Research: A Career at NASA Langley Research Center*. Washington, DC: National Aeronautics and Space Administration, 1998.

Phillipps, William Hewitt. *Journey into Space Research: Continuation of a Career at NASA Langley Research Center*. Washington, DC: National Aeronautics and Space Administration, 2005.

Powers, Sheryll Goecke. *Women in Flight Research at NASA Dryden Flight Research Center from 1946 through 1995*. Washington, DC: National Aeronautics and Space Administration, 1997.

Reginald, William. *The Road to Victory: A History of Hampton Roads Port of Embarkation in World War II*. Newport News, VA: City of Newport News, 1946.

Rice, Connie Park. *Our Monongalia: A History of African Americans in Monongalia County, West Virginia*. Terra Alta, WV: Headline Books, 1999.

Roland, Alex. *Model Research: The National Advisory Committee for Aeronautics, 1915–1958*. Washington, DC: National Aeronautics and Space Administration, 1985.

Rossiter, Margaret W. *Women Scientists in America: Before Affirmative Action 1940–1972*. Baltimore: Johns Hopkins University Press, 1995.

Rouse, Jr. Parke S. *The Good Old Days in Hampton and Newport News*. Petersburg, VA: Dietz Press, 2001.

Smith, Bob. *They Closed Their Schools: Prince Edward County, Virginia, 1951–1964*. Chapel Hill: University of North Carolina Press. 1965.

Sparrow, James T. *Warfare State: World War II Americans and the Age of Big Government*. New York: Oxford University Press, 2011.

Stillwell, Wendell H. *X-15 Research Results*. Washington, DC: National Aeronautics and Space Administration, 1964.

Warren, Wini. *Black Women Scientists in the United States*. Bloomington: Indiana University Press, 2000.

Wolfe, Tom. *The Right Stuff*. New York: Macmillan, 2004.

Woodbury, Margaret Claytor, and Ruth C. Marsh. *Virginia Kaleidoscope: The Claytor Family of Roanoke, and Some of Its Kinships, from First Families of Virginia and Their Former Slaves*. Ann Arbor, MI: Ruth C. Marsh, 1994.

Wright, Gavin. *Sharing the Prize: The Economics of the Civil Rights Revolution in the American South*. Cambridge, MA: Harvard University Press, 2013.

Articles

Bailey, Martha J., and William J. Collins. "The Wage Gains of African-American Women in the 1940s." National Bureau of Economic Research, 2004, http://www.nber.org/papers/w10621.pdf.

Branson, Herman. "The Role of the Negro College in the Preparation of Technical Personnel for the War Effort." *The Journal of Negro Education*, July 1942, 297–303.

Burgess, P. R. "Uncle Sam's Eagles Saved Hampton." *Richmond Times Dispatch*, January 13, 1935.

Collins, William J. Race. "Roosevelt and Wartime Production: Fair Employment in World War II Labor Markets." *American Economic Review* 91, no. 1 (March 2001): 272–86.

Dabney, Virginius. "To Lessen Race Friction." *Richmond Times-Dispatch*, November 13, 1943.

Darden, Christine M. "Affordable Supersonic Transport: Is it Near?" Japan Society for Aeronautical and Space Sciences lecture, Yokohama, Japan, October 9–11, 2002.

Davis, John A., and Cornelius Golightly. "Negro Employment in the Federal Government." *Phylon* 6, no. 4 (1945): 337–46.

Dunnigan, Alice A., "Two Women Help Chart the Way for the Astronauts." *Norfolk Journal and Guide*, July 6, 1963.

"Four Women 'Engineers' Begin Jobs." *Norfolk Journal and Guide*, May 22, 1943.

Frazier, Lisa. "Searching for Dorothy." *Washington Post*, May 7, 2000.

"Funeral Services Held for James F. Goble." *Norfolk Journal and Guide*, December 29, 1956.

Gainer, Mary E. and Robert C. Moyer. "Chasing Theory to the Edge of Space: The Development of the X-15 at NACA Langley Aeronautical Laboratory." *Quest Spaceflight Quarterly*, no. 2, 2012.

Goldstein, Richard. "Irene Morgan Kirklady, 90, Rights Pioneer, Dies." *New York Times*, August 1, 2007.

Gup, Ted. "The Ultimate Congressional Hideaway." *Washington Post*, May 31, 1992.

Hall, Jacqueline Dowd. "The Long Civil Rights Movement and the Political Uses of the Past." *The Journal of American History*, March 2005, 1233–63.

Hall, Phyllis A. "Crisis at Hampton Roads: The Problems of Wartime Congestion, 1942–1944." *The Virginia Magazine of History and Bibliography*, July 1993, 405–32.

Heinemann, Ronald L. "The Byrd Legacy: Integrity, Honesty, Lack of Imagination, Massive Resistance." *Richmond Times-Dispatch*, August 25, 2013.

Hine, Darlene Clark. "Black Professionals and Race Consciousness: Origins of the Civil Rights Movement, 1980–1950." *Journal of American History*, March 2003, 1279–94.

"Lady Mathematician Played Key Role in Glenn Space Flight." *Pittsburgh Courier*, March 10, 1962.

Lawrence, Dave. "Langley Engineer Is Remembered for Part in History." *Daily Press*, August 21, 1999.

Lewis, Shawn D. "She Lives with Wind Tunnels." *Ebony Magazine*, August 1977, 116.

Light, Jennifer S. "When Computers Were Women." *Technology and Culture*, July 1999, 455–83.

McCuiston, Fred. "The South's Negro Teaching Force." *The Journal of Negro Education*, April 1932, 16–24.

"Newsome Park to Open Soon; Shopping Center Is Feature." *Norfolk Journal and Guide*, March 27, 1943.

Reklaitis, Victor. "Hampton Archive: J. S. Darling: Leader of Seafood Industry in Hampton." *Daily Press*, August 27, 2006.

Rorty, James R. "Virginia's Creeping Desegregation: Force of the Inevitable." *Commentary Magazine*, July 1956.

Rouse, Parker. "Hampton Archive: Early Days at Langley Were Colorful." *Daily Press*, March 25, 1990.

Shloss, Leon. "Russia Said to Have Fastest Fighter Plane." *Norfolk Journal and Guide*, February 18, 1950.

St. John Erickson, Mark. "No Easy Journey." *Daily Press*, May 1, 2004.

Stradling, Richard. "Retired Engineer Remembers Segregated Langley." *Daily Press*, February 8, 1998.

Thompson, James G. "Should I Sacrifice to Live 'Half-American'?" *Pittsburgh Courier*, January 31, 1942.

Uher, Bill. "Tuskegee Airman Reunites with 'Best Plane in the World.'" NASA .gov, June 10, 2004.

"USO Secretary Weds Navy Man." *Norfolk Journal and Guide*, November 25, 1944.

Vaughn, Tyra M. "After Civil War, Black Businesses Flourished in Hampton Roads." *Daily Press*, February 14, 2010.

Walker, W. R. "Mimosa Crescent, Post-War Housing Project, Started." *Norfolk Journal and Guide*, July 15, 1944.

Watson, Denise M. "Lunch Counter Sit-ins: 50 Years Later." *Virginian-Pilot*, February 15, 2010.

Weaver, Robert C. "The Employment of the Negro in War Industries." *The Journal of Negro Education*, Summer 1943, 386–96.

"What's a War Boom Like?" *Business Week*, June 6, 1942, 22–32.

INDEX